THE
THIN
EDGE

THE
THIN
EDGE

Coast and Man in Crisis

Anne W. Simon

HARPER & ROW, PUBLISHERS

NEW YORK, HAGERSTOWN,

SAN FRANCISCO, LONDON

THE THIN EDGE: COAST AND MAN IN CRISIS. Copyright © 1978 by Anne W. Simon. All rights reserved. Printed in the United States of America. No part of this book may be used or reproduced in any manner whatsoever without written permission except in the case of brief quotations embodied in critical articles and reviews. For information address Harper & Row, Publishers, Inc., 10 East 53rd Street, New York, N.Y. 10022. Published simultaneously in Canada by Fitzhenry & Whiteside Limited, Toronto.

FIRST EDITION

Designed by Ginger Giles

Library of Congress Cataloging in Publication Data

Simon, Anne W
  The thin edge.
  Bibliography: p.
  Includes index.
  1. Coastal ecology. 2. Coastal zone management.
I. Title.
QH541.5.C65S55   1978   574.5'26   76–26253
ISBN 0–06–013890–4

78 79 80 81 10 9 8 7 6 5 4 3 2 1

for W. W.
with love

CONTENTS

FOREWORD

The temperature was over 100 degrees Fahrenheit the day we went to the North Sea beach near The Hague in Holland. In the unusually intense summer sun, crowds on the wide sand expanse were enjoying the breeze coming off the water; families picnicked, strolled. The special section for those who abjure bathing suits was not particularly noticed except by us Americans. But what we noticed even more was that, despite the cruel heat, no one was in the water.

At the tidemark we found out why. As waves rolled onto the sand, yellowish brown water foamed into masses of dirty bubbles, and in the shallows, endless numbers of lavender jellyfish, veined with purple, waited for prey, their tentacles extended in the viscous brown sea. Stranded on the beach, they became tinted blobs that stung our feet. Two small boys peered down from the jetty, looking for fish, although there were none. Otherwise the Dutch confined themselves to the upper

reaches of the beach as they have done since the big
North Sea oil find started producing oil and fouling this
stretch of shore. We were told that it was unlikely that we
could safely swim from any beach on the North Sea's
perimeter.

We yearned for the cool clear ocean back home. Except that it is, as I have come to know, the same ocean,
and the familiar, once-wild Martha's Vineyard shore we
cherish is at this moment also headed for destruction, if
of a different variety.

Four years ago I wrote a book about the Vineyard's
deterioration. What has been discovered since confirms
the idea implicit in that study of a single island; it requires us to consider any such coastal fragment, island
or mainland, as part of a whole—in fact makes it impossible not to—if we want to understand the present coast
and direct its future. This book takes advantage of the
new perspective and looks at the coast as an entity. It is
among the very first explorations of the one-coast concept.

This new vantage point is extraordinary, and terrify-
ing. Wherever we turn on the coast we confront the awful
destruction of its magnificent natural system. We can no
longer escape the results of years of short-sighted use
but must, for the first time ever, witness the dying coast
and wonder if we can still save it. Knowing what we know
makes ours the crucial generation.

My years of research brought me into contact with
several hundred members of this generation with special
knowledge of the coast or responsibility for particular
parts of it. Their understanding contributes to this
book's fabric. They have shown me the coast, talked to
me about it, explained their own views of it to me. I
appreciate the time and attention accorded me by these
men and women—fishermen, landowners, developers,

industrialists, town officials, governors, senators, cabinet officers, bureaucrats on many levels, professors, scientists, conservationists. Their universal willingness to talk, walk, drive over impossible terrain or ride out a drenching storm in an open boat to show me one last inlet reinforces the urgency of the coastal crisis. Rather than list each name, I hereby acknowledge the unique contribution each of these people has made and thank them for it, hoping that the net result will justify their generosity.

I am inordinately grateful for the sturdy support of Dorothy Olding, distinguished literary agent; the superior quality judgment and skill dedicated to this book by its editor, Frances Lindley; the wise, kind, always candid criticism and never-failing interest of my husband, Walter Werner. The subject moves so quickly and through so many disciplines that it requires a partnership of minds to encompass it. Thus my ultimate gratitude is to Eunice Whiting, expert researcher, whose rare selfless ability to share the search for concept, fact and detail enabled me to pursue this approach to the coast as an entity and made an invaluable contribution to it over the years.

ANNE W. SIMON

New York City
May, 1977

1
THE THIN EDGE

The coast, that bright thin edge of the continent where you can sit with your back to the crowds and gaze into seemingly infinite space, is now a theatre of discovery. On the seashore where terrestrial life began, we have to use all the wits man has developed to figure out how life can continue, how to design our complex, fast-moving, energy-consuming existence without destroying nature's system of life support in the process. It is a compelling adventure. Wherever it leads, neither man nor coast will be the same again.

Survival on land and in the sea depends on a functioning coast. The coast keeps us from drowning, maintaining the present global balance of one-third land, two-thirds water. It nurtures fish and shellfish, birds and plant life, as it nurtures the ocean, the essential source of a third of the world's oxygen, the largest source of its protein. Its multiple processes are arranged in dozens of natural systems with myriad parts, each neatly slotted into an operation as sophisticated as the latest computer, as intricate as a vast jigsaw puzzle. Its abilities are exquisite in their detail, awesome in their grand accomplishments.

Universally we yearn for the coast with an inexplicable need for its serene horizons, for the endless, timeless rhythm of waves on rough rocks or smooth beaches, for the amplitude—plenty of sand, water, seagulls, seaweed, a harvest of sea-worn pebbles and minute sea animals in every wave. Here where the sea is shallowest, land is lowest, rivers slowest, there is dynamic interchange between water and earth, a phenomenon often believed to make passions run higher, emotions keener, the sense of well-being quickened. We come closer to our primitive selves on the thin edge, at once nurtured and excited by it.

Ever-obliging, its generous compliance has provided poetry, joy, convenience and profit. But now we glimpse its deeper nature, stern, inflexible, firm in principle and in its limits. Innocent use has been cut short in the 1970s by what has been discovered about the coast, most of it in the past half century, much in the last decade, and by what is tantalizingly beyond our present vision. The existing information is unequivocal; the coast is different from any other place on earth and has different requirements. There is no man-made substitute for its manifold natural functions. We do not want to get along without a working coast and we now realize that we literally cannot get along without it.

Around the globe, coast functions falter under the encrustations of twentieth-century civilization. The east coast of the United States is vividly representative of any coast, anywhere, magnifying every coastal dilemma in its 28,000 miles of shoreline—coast, offshore islands, sounds, bays, rivers and creeks—stretching from Maine to Florida, from rigorous to tropical climate, the rocky northern shore testifying to its glacial past, the long stretches of wide southern beach, having escaped the glacier, relatively flat. Thirteen states have a slice of this coast under their domain and separate laws. It is heavily developed, industrialized, crowded, with hardly distinguishable towns wedged into the megalopolis solidifying between Boston and Washington, although there are, almost miraculously, still a few places—Down East Maine, some of Georgia's barrier islands—almost as they were when our ancestors first set eyes on their virgin marvels.

Ever since the last Ice Age gave way to a warming sea, the coast had been a magnificently productive system. Enormous trees, enormous quantities of fish and fertile black soil on the Maine coast amazed early explorers. The few remnants of the towering forests are remarkable today where they have survived, mementoes of a time when they covered the shore: ". . . goodly tall Firre, Spruce, Birch, Beech, Oke very great and good," says James Rosier, clerk on Sir George Waymouth's *Archangel*, sailing on a fair June day in 1605 past Monhegan Island, "a meane high land," to George's River "as it runneth up the maine very nigh forty miles toward the great mountains." Upon the hills, Rosier says, "notable high timber trees, masts for ships of 400 tons."

The adventurous men in their small ships were no less amazed by the waters teeming with huge fish of all varieties, a sight we can only imagine. "While we were at shore," Rosier relates, "our men aboard with a few hooks got about thirty great Cods and Haddocks, which gave us a taste of the great plenty of fish which we found afterward wheresoever we went upon the coast." The *Archangel* found "Whales, Scales, Cod very great, Haddocke great, Plaise, Thornbacke, Rockefish, Lobster great, Crabs, Muscles great with pearls in them, Tortoises, Oisters"—the list is long. Haddock and lobsters were so thick in the waters that some fishermen scooped them out with a bucket, salting them down in the hold for the long voyage home.

Even the much-traveled Captain John Smith was impressed. "Besides the greatness of the timber . . . the greatness of the fish and the moderate temper of the air," he writes, "who can but approve this a most excellent place for health and fertility?"

We can still see something of this excellent Maine coast. So too on a few remaining wilderness islands to the south, lush and semitropical, there are still wide sand beaches which sweep into a protecting line of dunes, while, behind them, gargantuan live oaks, pines and palms combine in a primeval forest, an enchanting world apart. When Giovanni da Verrazzano, still searching for the way to Cathay, explored this coast in 1524, wilderness was everywhere. He saw the beaches, dunes and estuaries that we struggle to keep fragments of. "The shoare," he says, "all covered with small sand, and so ascendath up for the space of 15 foote, rising in form of little hills . . . small rivers and armes of the sea washing the shoare on both sides as the coast lyeth."

It was a land "as pleasant and delectable as is possible to imagine." And on it, Verrazzano reported to his French sponsors, a delectable population, "people of color russet [who] go altogether naked except that they cover their privie parts with certain skins of beastes . . . which they fasten onto a narrow girdle made of grasse very artificially wrought, hanged about with tayls of divers other beastes."

In the millennia that man has inhabited the ancient edge of this continent, he has taken the short-term view of gratifying his desire for pleasure and security, shooting a few birds in the marsh for dinner, trapping furred animals, going fishing. Great piles of shells unearthed by archeologists, some charred, some halved, testify to the enjoyment of clambakes and clams on the half shell as early as 4000 B.C. Along with the shells in the middens, there are bones of otters, seals, whales, all sorts of fish, suggesting that their fur, fat and meat were used and valued. Coast dwellers took what they found when they wanted to, there being no apparent reason not to.

The coast's abundance welcomed colonists with the necessities—the fertile soil, wild game and fish, great timbers—that made settlement of the new world possible. The number of settlers was small, their requirements modest; the shore provided food and shelter and the water their only means of transportation. It was not until 1722 that for the first time a team of horses was driven from Connecticut to Rhode Island on a dirt path, winding through dense woods from one coastal clearing to the next. The coast, apparently constant and indestructible, continued to perform its functions.

It continued, in fact, through centuries of settlement and development, continued valiantly through industrialization. It continued, if somewhat less efficiently, as the vast coast population zoomed from 29.8 million people living within fifty miles of this narrow strip in 1940, to 48 million people in 1970, almost a quarter of all Americans, and the proportion increases three times as fast as the national average. All U.S. coasts together (including the Great Lakes) contain the nation's seven largest cities, account for 53 percent of its population and 90 percent of its population growth, and it is anticipated that, by the year 2000, two hundred millions will squeeze themselves into smaller and smaller segments of the thin edge.

Unprecedented numbers of people swarming onto the coast make unprecedented use of it in this technological age. The ravenous growth society devours many parts of the continent for its expansion requirements, but the thin edge, with its special attributes, is most delectable of all. It is a magnet for growth.

It is a magnet for people. On every coast the people business burgeons, lining the shore with stacked condominiums interspersed with mobile-home parks, marinas, dense second-home developments. A roaring tourist industry ricochets off coastal highways with its accompanying eateries, motels, neon-lighted putting greens. Coast recreation—surfing, swimming, snorkeling, sport fishing—escalating to an average ten days per year per American, is a profitable business.

The great bays of the east coast—Chesapeake, Delaware—and other estuaries are a magnet for heavy industry, refineries, power plants using water for cooling processes. Forty percent of the nation's industrial complexes edge its estuaries, 50 percent of its manufacturing facilities, and the east coast has more than its share. On the Delaware River, for example, utility companies plan 42 new power plants by 1986; one of these alone will evaporate 54 cubic feet of water per second, a loss equal to that of a small city. The activities of all these people and industries bring waste in unprecedented quantity to the shore, dumping it into the water.

Offshore waters are a magnet for the oil business. The colonists' lifeline becomes the tankers' trek as they ply and break up alongshore. Superships, with moorings approved for the Gulf coast in 1977, will soon rock in east coast offshore swells, and tracts of its Outer Continental Shelf have been leased to oil companies, now preparing to start full-scale oil fields in the Baltimore Canyon depths and in Georges Bank off Massachusetts, relatively shallow water warmed by the Gulf Stream and long a fabled fishing ground.

In the last ten years the coast's magnetic pull has become stronger than ever—more industry, more oil, more people, hotels, motels, boatels, more sewage, more waste . . . and more pressing evidence that the coast has limits, an idea hardly known and little considered until now. Sometimes quietly, sometimes violently, the coast is informing us that there is a saturation point beyond which its natural functions no longer flourish, often diminish, or simply cease.

The fastest-growing area in the United States is said to be the Florida Keys, a sixty-mile strip of islands and reefs some ten miles wide. At the present rate, the two millions who now crowd this reef will increase to ten millions before the century ends. Under the jammed Keys, reef-building corals, the only such colony in U.S. continental waters, are dead, their massive branches skeletons, covered with white spots where the organisms once grew. If you go snorkeling there, gliding past the dead coral mass, any fish you see, a dwindling population, are likely to be diseased and deformed. Biologists say coral requires warm, well-oxygenated water, that too much sewage and too much silt from dredging and filling for new buildings have suffocated the coral that built the Keys that are attracting humans faster than any other place in the country.

A coral reef, suffocated by the human life it supports, is a signal, quiet enough to go unremarked in the rush to cover the remaining inches of the Keys with concrete. Although such action can't further harm the coral, already smothered, it has other effects, noticeable wherever man transforms the soft sand shore into an inflexible wall.

The coast protects higher land by using wave and wind energy and gravity to build sand barriers that resist storms and pounding surf, as in the surprisingly sturdy barrier beaches that guard much of the east coast. In this era when the sea level is rising throughout the world, water encroaches on the shore and the coastline retreats. These barrier beaches reveal the remarkable ability to move inland along with the shore, rolling over on themselves to migrate with their entire ecosystem—beach, dune, marsh—intact. A North Carolina island has just performed such a giant somersault in less than a century. Pace is the key: the shore must move at its own speed, when and where it will. Interfere with its pace and it will neither guard nor turn somersaults. Before this life-saving information was discovered, much of the shore had been covered with mammoth concrete development, preventing free movement of sand and water, a matter of considerable concern. It will be of more concern if the hurricane cycle, which has been in an unusual and seductive lull during the 1970s, years of the most concentrated seaside building, returns as expected, roaring along the concretized and thus dangerously vulnerable Atlantic coast. "The cost in dollars and lives of the next Camille-size hurricane will be staggering," a scientist predicts.

Behind the barriers where rivers empty into protected bays, the coast manufactures food for marine life by a mix of fresh and salt water, wetland grasses, sun and tide, delivers it to coastal species, for many of which these sheltered spaces are a necessary habitat during part or all of their lives. It has recently been found that, acre for acre, wetlands are the most productive land on earth. Without protective barriers they will drown. Already, thousands of such acres along the eastern seaboard have been irretrievably lost; they were filled in, converted to high land, dredged or otherwise stressed, before their value became known, and even after.

The man-coast love-hate relation changes with each discovery of a new facet of coastal character. We begin to see limits beyond which the coast cannot function, where its nurturing nature turns hostile, antagonistic to life, suffocating, drowning, poisoning. The signals are ever stronger, ever deadlier.

The Glorious Fourth weekend of the nation's Bicentennial, when New York was momentarily a festival city, applauding the muster of the tall ships in its rivers and harbors, the skipper of the *Faye Joan,* trawling for whiting off New Jersey, winched in his net, spilled a thousand pounds of fish on the deck. "The contents stank," David Bulloch, an observer, says. "The fish were dead, a few dying, most decaying. The crew worked, barely breathing, to shovel the fish over the side. The urge to vomit was overpowering." Diving to investigate, Bulloch and others found unusual dark brown water and below it, on the cold bottom, piles of dead fish, crabs, lobsters, mussels, "a foot-thick black mass of decay swaying with the surge of the sea." By August the killing sea extended for three thousand square miles. "Everything was dead," a microbiologist on the scene reports. "Nobody can remember anything like this." The level of dissolved oxygen in the water, required to sustain marine life, fell to zero, overpowered by the torrents from the rehabilitating city's sewers.

A different coast from the natural shore explorers found, different from the settled coast of the start of this century, different, even, from what it was just a decade or two ago when we apparently passed many of its limits without even knowing it! Each day we venture further into the unknown character of a world without a working coast, to date the only generation to experience such terror. But each day we are better equipped to stake out the limits for man on the coast. We begin to decode the signals that issue from the thin edge.

Each change of one part of the coast system affects

other parts. Some connections have been discovered, such as the linkage of barriers and wetlands; some are still unknown. It may be that the shore is so complex that we will never completely quantify the results of a change, that we must always play Russian roulette with our coastal intrusions. Consider the mid-seventies decision to explore for oil in Georges Bank. The results circle out like the ripples from a pebble thrown into a quiet pond, with no end in sight. One such ripple catches up coastal flora and fauna in an interconnection never imagined.

Marine biologists, of late particularly interested in the common seaweed kelp because it can be cleanly, cheaply converted to fuel, are surprised to find bald areas in underwater kelp forests off the northeast coast. More than eight times the expected number of sea urchins in great herds are grazing the kelp down to bed rock, John Culliney says in *The Forests of the Sea*. Some suspected increased sewage in the waters, enjoyed by the prickly half-sphere animals, might be responsible; others believed that overharvesting lobsters, the urchins' most avid predators, could account for the multiplying urchin armies and the vanishing kelp.

There are fewer lobsters to eat urchins and to be eaten by man for a reason that these strange succulent creatures have long kept hidden. Only in 1970 was it discovered that some lobsters, primitive, awkward and slow-moving as they may seem, have each fall for thousands of years walked 150 to 200 miles across the rock and sand bottom of the offshore sea to Georges Bank, enjoying the winter in the nonfreezing temperatures there, walked back again in spring to coastal waters to copulate under sheltering rocks where, in a miracle of precise timing, the females shed their hard shells to make it possible for the waiting males to enter their bodies and deposit sperm.

Something new has happened in the underwater world of Georges Bank, a change so fascinating to lobsters that they hang around like hooked junkies, Culliney says, the vernal journey back to shore and its primal purpose forgotten. There is oil in these waters now, oil from exploratory digs, oil from tanker spills, more oil than ever before, and the lobsters, it has just been found out, are mightily attracted to it, will attack and eat kerosene-soaked paper in laboratory tanks, seek it out in their wintering grounds. If Georges Bank oil wells start in earnest, propagation in the wild of the migratory branch of *Homarus americanus* may be over forever.

As a single change of balance it could be inconsequential; sea urchins are unlikely to take over the world, lobsters can, perhaps, be successfully cultivated. But as representative of countless changes, revealed or still unknown, the kelp-urchin-oil-lobster cycle is grave and deeply troubling.

The change is an archetype coast module, the module that appears in hundreds of fragments and forms, in unexpected places with sometimes inconvenient, sometimes punishing, sometimes murderous effect. The more such a module is pieced together, the clearer it becomes, suggesting as it does that the essential coast character is its intricate, indisputable interconnection. Discovery of the coast's amazing systems advances our knowledge of this interlocking nature of the thin edge where we stand, precariously, listening to its silent scream.

2
SAND

The quintessence of the coast is restlessness. It is forever moving, changing its shape. Some changes happen overnight—a sandy beach is covered with rocks, a sand bar appears where there was none, a dependable bank tumbles into the sea. Some take longer; the slowly inundated shore loses a foot or two to the rising sea each year, even more in some places. The slowest change has been going on since the first rock was exposed to the first water. Over the ages, rock disintegrates to sand. You can't see it happen in a mere lifetime but on the east coast you can see the before and after, the ragged, rough, relatively young rocky shore in the north, the smoothed out, curving, flattened sandy beaches in the south.

The absolute quality of coastal change is better known now than it ever has been. With great care we peel back one layer after another of geologic time to know how the coast was formed, how its kinetic structure, its very genes, came into being. Despite this knowledge we try to alter the restlessness, to make the shore a static place. In doing so we interfere with the sand system, one of the planet's ancient processes, stirring up trouble beyond belief.

Motion lies behind the sand system. It is free and unhampered on Cumberland, a Georgia barrier island, where almost nineteen miles of beach sweep along the Altantic, virtually in its virgin state. The sand's characteristics, untouched, mark it as an ancient habitat. Sharp claws of giant sea turtles indent the sand, making a track where this prehistoric species, some weighing nine hundred pounds, still lumber out of the sea and up the beach to lay their eggs out of reach of the tide, a hundred or more eggs, each the size of a ping-pong ball. One egg may lie exposed, its mottled gray shell broken by a marauding wild pig or raccoon, its yolk, as yet undevoured, quivering in the sun-baked sand. Horseshoe crabs whose ancestors lived with dinosaurs still sidle through the shallows. Clams, mussels, oysters, unchanged since their evolution, shorebirds whose skeletons have not varied through the millennia, still live out their primitive life span on Cumberland's beach. Its sand is their environment.

Sand meets water's force with its natural tendency to move; its soft answer turns away the sea's wrath. Waves energized by the sun, moon and wind crash on the wide expanse as they have done every day and night since the island was formed. The use of sand to receive and repel this merciless attack is a triumph of natural engineering. The smallest grain of sand is almost indestructible, especially when wet, keeping a film of water about itself by capillary action. Because of this liquid cushion, there is little further attrition. Even a heavy surf cannot cause one sand grain to rub against another.

Under water, sand swirls with currents, shaping the

shoreline, erecting barriers, building beaches. On a calm day you can splash out to a sand bar off Cumberland, perhaps submerged today although tomorrow it may surface; visible or not, it exhausts some of the waves' energy. The remaining force beats against the sand that slopes up to the shore and against the beach itself, some hundred yards wide at low tide, where it is finally dissipated.

Cumberland's most spectacular sand defenses are the dunes—sand waves—blown up from the beach by prevailing winds into a double line running parallel to the shore, last and sturdiest protection against waves and storm. The seaward line of shifting dunes moves an unbelievable ten to fifteen feet a year. Where they have overtaken and smothered a forest, twisting tree skeletons make a stark surrealist edge to the beach; depending on the light, it may be shining silver, a black silhouette, or pink-dappled, reflecting the beach sunset.

Eventually the dunes are caught and held stationary by plants specially designed to survive sun, wind and salt spray in harsh combination, able to root and flourish on the hot, dry sand. Some are tough, thin blades like beach grass, which responds to the breeze by tracing its perfect circles in the sand while its roots grow deep and fast, searching for water in the dune's innards where they join each other, making a reinforcing network. Some spread wide and flat to catch moisture, like the rose and green circles of seaside spurge; some are wax-coated for protection from the salt. On Cumberland, sea oats, beach peas and stunted goldenrod grow at the edge of the dunes; beyond them, yucca and prickly pear, cabbage palms and the saw palmetto are reminders of the close-to-tropical climate, as are low-growing live oaks, draped with moss, growing on the back of the dunes. Along with pines and cedars they eventually make the dune a rooted hill.

Three dunes on the string of Georgia barrier islands have reached fifty feet, as tall as any on the east coast, and all three are on Cumberland. On one, red cedars have been shaped by the blowing sand and wind to form a smooth canopy as perfect as if pruned by expert hands in an Edwardian garden. Behind it, in a fresh-water pond (locally known as a slough) you can glimpse the swish of a great alligator moving through the tangle of yellow water lilies which flourish there.

All of it—beach, dunes, sand-growing plants, ponds, and seaside forest—is part of the sand system. Offshore sand is its matrix. For centuries the vast flowing under-water sand bed, energized by littoral (alongshore) currents, has supplied sand to the beach; for even longer it has been the way station for sand coming to the sea from rocky mountains and rock coasts.

Each part of the system is now threatened by the descent upon it of our gargantuan, coast-oriented, recreation-minded, industrialized society. For us, shifting sands is a convenient cliché but an inconvenient reality. We do not accept it. We call it erosion and engage the United States Army Corps of Engineers to fight it. "Our campaign," the Corps says, "must be waged with the same care we would take against any other enemy threatening our boundaries." In the 1970s it found that erosion was at a "critical stage" on 2700 miles of the U.S. shoreline, two-thirds of it on the east coast, where it would cost $1.1 billion to halt it. The Corps gives out some frightening figures on erosion; a foot a day in North Carolina, four feet a year on New York's Long Island, sixty feet a year in Maryland. It recommends structural defenses for stabilization, supports some twenty bills in Congress to strengthen beach erosion laws.

Our battle with this enemy is new, new with the crowds and the diminishing supply of waterfront. It is new since recreation became the leading economic activity of the coast, increasing by a staggering 76 percent in the past fifteen years and accounting for more than $15 billion spent in a single year, new thanks to a technology that lets us believe we might actually change the coast's character to suit ourselves. We think we can satisfy people who want to swim from a sandy beach and a sandy bottom and at the same time want to live directly on the water, building in solid lines along the shore. A catalogue of our intervention with the coast's structure—harbor jetties, bulkheads, seawalls, breakwaters and the rest—describes them as powerfully erosive. A National Geodetic survey suggests that our overwhelming presence on the coast may be tilting the crust seaward, hastening the rise of waters . . . erosion.

It is the first time man has undertaken to second-guess the natural business of sand, an awesome assignment. The mobility we now interfere with started in some shadowy time when gas became mass and the young earth issued molten rock from its core to form a solid crust. For billions of years the crust was never still. Rock was tossed into mountain peaks only to be flattened on the ocean floor or brought down again into the planet's center to reform in new chemical combinations and reappear on the crust. Each enormous move took eons, time beyond our comprehension. The present continents only started to take shape 75 million years ago, shifted, reformed and were further sculpted by the Ice Age that started a million or more years back, ended twelve to fifteen thousand years ago—or perhaps just receded for one of its interglacial periods, Charlton Ogburn suggests in *The Winter Beach*. Ice from the Arctic regions could come back at any time, he says, finding it easier to bear "the monumental ugliness we have wrought" by thinking that the earth might cover it all over with ice and start afresh.

Ice sheets a mile high in places covered the northern half of the continent, melting and freezing again at least four different times. In the south, this moved the shore-line, revealing land in successively lower levels. Georgia's shore was protected by barrier islands then as now; there are five rows of land-locked islands, described by people in the neighborhood as bluffs. One investigator, canoeing between the fragrant banks of a coastal river, came upon stunning testimony—a forty-foot golden sand dune standing in the dark green forest.

In the north, glaciers ground down mountains, rolled rock and debris across the landscape. When the ice melted, released water drowned what had been exposed. The thousands of islands off the Maine coast are the tops of drowned rocky hills, the long peninsulas reaching into the bay are drowned mountain ridges, both now covered with dark canopies of spruce, which by contrast make the sparkling blue water even brighter. "This," one enthusiast writes, "is not so much *a* drowned coast as it is *the* drowned coast of the world."

It will not last. Despite the comfort of the well-loved hymn, there is no eternal Rock of Ages, even on the Maine coast. Hard, dense, solid though it may be, it will disintegrate, if the planet survives long enough, and become what geologists call a "mature coast," smoothed and sandy. Water crashing against a rock's edges, freezing and expanding in its crevices, raining on it through the millennia, grinds it down. Lichen that spreads its delicate green circles against the rough gray granite secretes a chemical that helps erosion; dead plant matter performs another such service. Little by little, softer rock faster than harder, older before younger, rock crumbles.

The end product, rock's irreducible minimum, is sand. Hold it in your hands and you are in touch with the planet's essence. Each grain has been part of the earth's solid crust at one time or another, eventually to be freed from rock to exist as a grain again, its particular chemical structure intact. It might be a quartz crystal, said to be the most widely disbursed mineral on earth, or a tiny garnet; a grain of clear feldspar, so hard it cannot be scratched with a knife, or mica chips, brown, black or white. Although the tiny particles might be compressed again, perhaps into sandstone or slate, they will keep their identity to return at some future time as grains of sand.

Sand travels, the heavier grains by water, the lighter —more rounded and often frosted by sandblast—by air. It blows from eroding mountains or is carried by brooks and rivers down to the sea, adding to sand from boulders grinding in the surf, from rocks on the shore. Inland, the local sand pit is the descendant of deposits left by the glacier on the continental shelf, a wide, flat ribbon which runs along the coast, above the sea and under it, until, some miles out, it slopes down to the ocean depths. The great underwater sand fields on this shelf are glacial descendants too, renewed over centuries by continuing extraction of sand from rock and transportation of it into the ocean. The final stage of nature's meticulous plan of freeing and moving sand, a fluid system as old as the earth, is that soft, warm, rippling edge of land which we prostrate ourselves on, run our fingers through, build into castles at the edge of the shore . . . and into the requirements of our society.

Plain simple sand, as ordinary a part of the earth as any, unnoticed, unsung, unremarkable, an item to be shaken out of shoes, to fill the children's sand box, to mix into concrete, occasionally to have its picture taken when freshly patterned by an outgoing tide. How exceptional its presence now becomes.

We use 944 million tons of sand and gravel a year, 90 percent of it to build buildings and pave roads—four times as much as twenty years ago and only half our estimated need in 1980. Until now there has been enough sand in the glacial deposits near the great coastal cities where most building takes place. But now these pits are being exhausted. Land close to cities is becoming too valuable to use for sand pits and transporting sand from pits farther away adds to its cost. The only remaining nearby source is under the water.

For the first time we look toward offshore sand to meet the requirements of nonstop growth. Jamaica Bay, which supplied enough fish to support Long Island's first known settlement in 1636, yielded 53 million cubic yards of sand to build Idlewild Airport in the 1940s. An expert dredging technology has developed; a recent improvement, invented by the Japanese, is a radio-controlled, automatic amphibious digging device called the UA03 hydraulic excavator which has precise control of submerged boom, arm, and bucket movements. By 1972, such super-machines made it possible to remove 10 percent of the annual sand consumption, almost 100 million tons, from under the water with magical efficiency.

There is a lot of sand lying off the east coast. No one is sure how much. The Army Engineers, who, as we will see, have reason to want the amount to be enormous, estimate it at 450 billion tons. Even if there is only half that much, why worry about removing a billion or two tons a year?

Sand is listed in most coastal manuals as an "Unrenewable Resource." Whatever proportion of it we choose to lock into concrete is a final decision, at least for the next century or two (like rock, concrete won't last forever), a decision with significance for the coast which we are just beginning to understand.

We have already created a sand shortage. And a shortage of offshore sand discomforts and deprives people who live on the beach. Some try man-made traps for sand such as jetties or groins (barrier-type structures) in the water adjacent to the beach. This interrupts the normal alongshore sand flow so that other beaches as well become sandless. "Since less and less sand is available as a natural supply," the U.S. Corps of Engineers says, "it is frequently necessary to place sand artificially to fill the area between the groins." Wherever *that* sand came from in turn suffers a shortage.

The shortage shows up *in extremis* on what was a mangrove-covered barrier island somewhat to the south of Cumberland and, like it, once a sand reef, building its natural defenses against the sea, growing by means of its mangroves, which root in shallow water to catch mud and silt. Named for its beach, this reef—Miami Beach— ironically is now virtually beachless. Man's determination to get to the water and dig his toes into the sand has made it impossible for him to do just that, at least on the island said to be the most densely concentrated luxury resort area in the world.

Hotels massed in a solid front as close to the beach as possible have transformed the protecting line of dunes into a concrete wall. The beach, thus deprived, is unable to hold the littoral sand drift that comes to it from the north and is eroding fast, helped by storm waves bouncing off hotel bulkheads. Already hurricanes have flooded hotel lobbies and deposited two to three feet of sand in the streets. Damage is much greater than in the years when there were dunes to give the soft answer to autumn hurricanes. The problem is unlikely to go away. Even during the present lull, hurricanes head for the beachless beach on Florida's tip with recorded regularity, one or two coming uncomfortably close every year, twenty-five of them having hit squarely and savagely since 1830.

Hotel owners, who once turned sand into dollars as fast as the suntan lotion could flow, and Florida, which benefited from such sand-derived profits, and people who won't settle for chlorine-heavy pools but like to walk, sit or sleep on the warm sand, want the beach back.

Again the appealing solution is offshore sand. The Corps of Engineers, not known for its profound understanding of natural resources, proposes to dredge for enough sand to build ten miles of beach, two hundred feet wide, with a two-and-a-half foot dune at its inland edge. The hotel owners oppose the Corps's proposal because the use of public funds would deposit a swarming public beach at their billion-dollar doorsteps. Their counter-proposal, however, is also based on taking sand from offshore.

Troubles that plague the crowded little Florida island exist one way or another wherever man has interfered with the sand system. Trample down dunes, breaking the delicate network of plants and grasses, prevent the free flow of the littoral current, diminish the sand supply beyond its tolerance . . . sand stops protecting us and, if hard-pressed enough, goes away.

What we do not know is what else may happen if we scoop more and more sand from offshore beds. It could wreak havoc with the ocean's food chain on which we depend. Changing the composition of the sea floor from sand to mud or rock will change what lives there.

Inhabitants of sandy bottoms, being ancient life forms, have become highly particular about where they can survive, and have developed specialized organs to suit their environment. Eugene P. Odum, author of *Fundamentals of Ecology,* bible of ecologists, compares the inhabitants of a sand and a rock bottom: "Very few of the conspicuous dominants are common to both." Some species are specialized to the bottom of the sea, some to a certain depth; even seaweeds (actually larger forms of algae) grow green, brown and red from shallow to deeper water, in that order. Flounder and other flatfish feed off a shallow sandy bottom, as do herring, sardines and anchovies. Take away the sandy bottom and there are strong indications that the population will move out.

When we remove sand from beaches, we also meddle with the extraordinary life styles of shore species, much more precisely adapted to the environment than man, much less capable of picking up and living elsewhere. Beach life is different from rock life but both these seaside habitats are zoned more strictly and democratically than the shores of America are zoned for people. There's a place for all that need to live there.

Beach dwellers living in the zone farthest from the water are described as land types by naturalists Mildred and John Teal, who spent four years peering at the inhabitants of Georgia's Sapelo Island. Ghost crabs for example, make dunes their home, burrow in the damp sand during the day, look for food at night. They dip in the surf to wet their gills and are on the way to becoming fully land-based, the Teals say, as their gill cavity develops into an air-breathing lung. The second zone, occasionally immersed by incoming waves, is inhabited by a variety of worm burrowers that exist on algae and dead plant matter left on the beach by the tide, and by certain fast diggers like the mole crab, which can bury itself in a second or two, leaving just its mouth and feathery antennae exposed to catch plankton when the waves come in. Closest to the water are carnivorous snails and worms, able to feast off animals plentiful at the edge of the sea, burrowing to keep themselves in place against the tide.

A weak-muscled little ghost shrimp with rudimentary eyes secures its burrows by meticulously secreting a cementlike substance to hold the walls erect. It rarely comes out except to deposit its fecal pellets around the burrow's tiny opening, is so well-adapted to its place on the beach that it has lived there since the last Ice Age, faithfully building a line of cement burrows wherever the shore happened to be. The burrows remain today to delight geologists with their mute, precise information. Long before the shrimp came into being, soft-bodied worms, some more than a foot long, made burrows on beaches that existed 620 million years ago off North Carolina. Discovered and carbon-dated in 1975, these beach creatures become the oldest known animals in the United States.

Would the near-blind little shrimp survive a sandless place or just drop out of sight, leaving future civilizations bereft of its small cement markers? Does it matter?

We are gambling that its loss would be inconsequential or that, by some miracle, sand-specialized creatures from ghost shrimp to mole crab will oblige us by adapting themselves to less and less sand, an adjustment most of us are unable or unwilling to undertake for ourselves. A Long Island Planning Board rated its knowledge of coast stabilization—understanding the results of moving sand—"Poor." A 1972 coastal-zone conference discusses dredging. "There is not enough data to predict the rate of bottom organism recovery or reestablishment of natural bottom configurations," it reports. Britain, which has been mining sand off its east and south coasts for forty years, also apparently avoids the question. So far, the impact of offshore dredging on British bottom dwellers and beach-adapted species "has not yet been evaluated."

Sand challenges what we do to the shore in many ways. It moves real estate around. Hurricane Agnes created a five-hundred-acre fan of islands in the mouth of the Susquehanna River in two days, a process estimated to take a century by normal accretion. Suddenly, there were the islands; to whom did they belong? We have not figured out the real-estate angles of such Acts of God, any more than we know who can be held responsible when beaches disappear. In these days of wildly escalating values of shorefront land, sand's age-old proclivity to move becomes a public dilemma. "The government owes us something for having paid taxes on that beach all these years," say the citizens of Maine's Popham Beach, which has more than half disappeared in the last years, while on St. Simons, a developed island near Cumberland, Georgia is suing a developer trying to build a hotel complex on newly accreted land, asserting that such natural beneficence belongs to the state.

Sand motion that started when the earth began is not likely to stop, whatever we may invent to try to stop it. But our increasing interference with its system alters its performance. Free-flowing sand has become a rare sight on the east coast. There are few virgin offshore sand beds that have not caught some wandering engineer eye, few Cumberlands that have escaped the sea walls, groins and jetties of development, few beaches where species, including ours, can continue their life patterns unimpeded, few dunes that are not trampled or built on, leaving a path for floods.

The sand system will not stand still for us. It cannot adapt to what we want from it. It is rebelling, and we are the losers. This very essence of the planet is trying to tell us something, and the message is not too different from that of other elements of the shore.

3
SEA

There is an extraordinary mix where land meets sea, a watery amalgam of civilization and the primordial ocean, mother of us all. These coastal waters are the link between man and the sea. The end product of what we terrestrial beings do eventually goes into them, transported there by the great rivers that pierce the coastline, by the water that washes the shore, or by man himself, deliberately or accidentally.

Until now we have depended on the vastness, the limitless ocean, the marvelous wild waters, believing the sea to be a resource without end, all-accepting, all-forgiving. "Roll on, thou deep and dark blue ocean, roll!" Lord Byron writes. "Ten thousand fleets sweep over thee in vain;/Man marks the earth with ruin,—his control/Stops with the shore." Ever since the ancient Greeks pictured Zeus in charge of whatever lay beyond the ocean rim, we have inched along in understanding it, tortuously charting its surface, measuring its depths. We learned that there were no Elysian fields beyond the horizon, but water, covering 70 percent of the globe, 330 million cubic miles of it as presently estimated, deep, continuous, salty. The more we measured, the more we trusted the enormous ever-rocking sea to be impervious to man. Our tampering has come about so fast that in 1950, just a second ago in ocean time, Rachel Carson didn't even suspect its effects. Her research led her to echo the same belief that inspired Byron. Man confronts his mother sea only on her own terms, Carson said. "He cannot control or change the ocean as, in his brief tenancy on earth, he has subdued and plundered the continents." By a curious twist, the very excellence of her work may have encouraged what she was sure could not happen.

We have discovered since then that we can and do control the ocean from the shore, change its capacity to support life on which our own lives depend. There is marine life in every part of the sea but it thins out in the dark depths where the sun cannot penetrate. Most of it teems in the surface layers, the top 2 or 3 percent of the ocean. Of these top waters, only a small part produce food and shelter for the rest, some where the tide runs over the land creating wetlands, some where the shallow sea surrounds the continents, a fraction of a percent of the ocean . . . the crucial fraction.

It is a terrible irony that at the precise moment of discovering that we are affecting the entire ocean from this crucial fraction, we are pressed to invade these waters in ways never possible before this era, and for urgent contemporary reasons.

There are 108 million submerged acres off the Atlantic coast, from shoreline to 200 meters of depth. Energy resources are buried there—oil and natural gas—that we want now as never before; there are, as has been seen, huge sand beds and fish and shellfish we are desperate to retrieve. There are also unprecedented crowds on the coast, producing proliferating tons of sewage emptied into this same fraction of the ocean. Corroding acids and hot industrial wastes are discharged into it, and barring an unlikely policy reversal, there soon will be some sort of supertanker ports—a string of man-made offshore islands is the favored proposal. "Whichever way the problem is solved," Noël Mostert says in *Supership*, "what cannot be solved is the damage any form of tanker unloading causes." The effect of oil on the waters is "the essential nightmare."

It is painful to absorb the fact that one generation, our generation, has not only poisoned the air to a degree which is causing us serious disabilities, and plundered the earth, but now is beyond any doubt corrupting the ocean. The mysterious human bond with the great seas that poets write about has a physiological base in our veins and in every living thing, where runs fluid of the same saline proportions as ocean water. It is no wonder that we are instinctively drawn to the sea and would avoid any sign that we are damaging it.

The craving to believe that the great seas will stay pure is powerful. Despite confirmed reports of widespread deterioration from changes we have caused in the crucial offshore waters, there is near-blind determination to equate seas and purity. Stores that dispense Health Food can't keep sea products in stock. There's a rush on sea salt, seaweed shampoo, on a kelp health capsule which, with a doubly sardonic twist, is said to relieve the effect on the body of car fumes.

The beginnings of the sea followed long after the initial explosion that filled all space when the temperature was about a hundred thousand million degrees Centigrade, Harvard professor Steven Weinberg says in *The First Three Minutes,* describing what in the last decade has become a widely accepted scientific "standard model" of the Creation. The density of the cosmic soup was four thousand million times that of water, Weinberg tells us; as the explosion began to cool, the temperature dropped, matter became less dense, ultimately to form the present universe.

Eventually solid spheres emerged from this primordial soup. One of the smaller was earth, 4600 million years old by present reckoning. The spinning rock emitted gasses from its molten core as it whirled through space; the gasses formed its unique atmosphere, which, when the rock cooled enough, condensed to form its unique ocean (all other planets being too hot, too cold or too small to be so favored). In this ocean, life began.

Recent theories of the event that no one was around to see agree that the sea then constituted a mix of the simple chemicals of the young earth that had been washed into it. Exposed to sunlight, some elements combined into larger floating molecules. A particularly remarkable molecule, formed by "an exceedingly improbable accident," Richard Dawkins says in *The Selfish Gene,* had the extraordinary property of being able to create copies of itself, the *Replicator,* as Dawkins calls it. Its copies soon spread throughout the sea. Replicators made copying "mistakes," some more stable and long-lived than others, and eventually discovered how to build survival machines for themselves that became progressively more elaborate. "Whether we call the early replicators living or not," Dawkins says, "they were the ancestors of life. . . . Four thousand million years later . . . they go by the name of genes, and we [all living things] are their survival machines." Some early cells became capable of photosynthesis, transforming the

sun's energy into food and supplying the oxygen that
would form our atmosphere. These plant cells were food
for the first animal cells. Both flourished in the warm,
rocking sea, multiplying into a variety of marine flora
and fauna, some of which would eventually find their way
onto land.

A comprehensible timetable of this version of prehis-
tory is to set it, as one interpreter does, against a year of
our time. January is about 4600 million years ago when
the earth was formed. Somewhere between March and
June the ocean came into being, its replicators and survival machines in July and August, dinosaurs in mid-
December. The ascent of man from animals began, according to this calculation, at about 10:30 P.M. on the
365th day.

Quite some minutes later by this calendar, the urge
in Europeans to find their way across the sea to the riches
of the Indies welled up till it burst in the mind of Queen
Isabella and the remarkable Genoese mariner. In *The
European Discovery of America,* Samuel Morison describes
the surprisingly fast trip of the three small square-rigged
vessels across the Atlantic at the northern edge of the
trade winds, the mutinies, the false landfalls, the determination of the Captain General, his enjoyment of the
air, which was, he recorded in his journal, "like April in
Andalusia." The night of October 11, 1492, was clear
and beautiful but the sea, Morison says, was the roughest
of the entire passage. Before dawn the *Pinta*'s captain
sighted a white cliff, shining in the moonlight. The little
fleet anchored, awaiting the day, then Columbus went
ashore in the *Santa María*'s longboat. Kneeling on the
gleaming coral beach, he thanked Our Lord, wept tears
of joy for having reached land, rose and named the island
San Salvador—Holy Saviour.

It was heroic, demanding, and madly adventurous to

cross the Atlantic during the century of exploration that
followed that first fateful voyage. But the lure of lands
filled with gold and spices had dozens of captains per-
suading their several sovereigns or others to back voy-
ages, chancing the wild winds and uncharted waters to
find the celebrated straits that weren't there. They discovered Nova Scotia and Newfoundland, where Norse-
men had come some 500 years before; they found Cuba,
Jamaica, Florida. A generation after Columbus, there
was still a blank in between, the east coast of the future
United States of America, waiting to be found.

The blank was filled in by Giovanni da Verrazzano,
gentleman explorer, who sailed the unaccompanied hun-
dred-ton *La Dauphine* with a crew of fifty—the maritime
mob, Verrazzano called them—to a landfall in the
Carolinas, down and then up the Atlantic coast, his habit
of anchoring well offshore causing him to miss many
important harbors and rivers. He did discover "a very
pleasant place situated among certaine little steape hills;
from amidst which hills there ranne down into the sea an
exceeding great streme of water," and anchored there,
protected from the winds, the first European to gaze on
New York Harbor. Today that place is less pleasant.
Spanned by a soaring, silvered, twentieth-century engineering triumph named for him (minus one "z"), the
Verrazano Bridge, what the explorer found there—
"roses, violets, lillies and many sortes of herbes, and
sweet and odiferous flowers different from ours"—have
long since given way to the first great assemblage of
concrete towers the world has known, and the harbor is
widely reputed to contain the world's most virulent waters.

Soon fishing fleets from England, France, Portugal
braved the turbulent seas to newly discovered fishing
grounds off Newfoundland and Cape Cod. It is said that

these sixteenth-century sailors, tossing through the winter westerlies of the North Atlantic to get to the Grand Banks in early spring and again in the summer were the best sailors in history, dependent only on the wind and common sense to get them there and back. By 1550, more than a thousand ships would fish over the Banks, their crews at first salting down the abundant cod, haddock and mackerel for the journey home, and later camping on shore to dry the fish in the sun.

The sea was a highway in those days, and later, colonists depended on it. They needed it to communicate with each other, to trade along the coast and with the home country, exchanging wood, furs, fish for manufactured articles, and they needed it as a source of food. Their first ship, the thirty-ton *Virginia,* was launched in 1607 on the Kennebec River in Maine; a century later, New England yards were sending a ship a day down the ways.

With a whole continent to settle, there was no reason to discover any more secrets of the sea than those that would take Americans where they wanted to go as fast and as safely as possible, and to find, as people still try to, the best fishing. Early colonists enjoyed, indeed counted on, the incredible bounty of the Grand Banks and Georges Bank off Cape Cod; cod made them rich. Why would they try to crack the secret of the fabled fertile fishing grounds, as long as the banks continued fertile?

Why would Columbus, with the Indies on his mind, investigate the phenomenon of the masses of brown seaweed encountered in the clear warm blue Atlantic between the Azores and the Bahamas? "Saw plenty weed" was his laconic journal entry for days, while his sailors, naming it sargassum after a native Portuguese rockrose, thought it might tangle in the rudder and sink the ships, a fear that survived as late as 1920, when steamers were still warned to avoid the Sargasso Sea.

It took thousands of bits of information, slowly collected, and entirely new and different motives, to unfold such mysteries. It took a Benjamin Franklin, ever curious, to inquire why American ships crossed the Atlantic two weeks faster than English ships and to find out from a Nantucket whaling captain that the Yankees were taking advantage of a current running eastward across the ocean at some three miles an hour, avoiding said current on their return trips. Subsequently Franklin published a chart of what he named the Gulf Stream, first ocean current to be mapped. It runs up the east coast of the United States, warming coastal waters till its course turns to cross the Atlantic. Since Franklin's time, knowledge of currents has advanced far enough to suggest that we might learn to control the weather by changing their course.

Now, almost five centuries after Columbus, having mapped the continents and touched the moon, we desperately need to understand the workings of what has become our last frontier. How little we know about the sea! Francis Shepard, who has had a distinguished fifty-year career in oceanography, says it was only in the 1960s that geologists had sufficiently explored the sea to be able to discard established ideas and take up "what seemed a prophetic dream of continents splitting apart and new oceans forming. We are in the midst of a great detective story," he says.

The sea is a hospitable place because its water is a total environment. An effective heat reservoir, it regulates its own temperature and that of the air, moderating extremes. With some tragic exceptions now becoming known, it can dissolve almost anything, and provides solutions of oxygen and carbon dioxide for its animals and plants, along with every nutrient mineral needed for life and some that are not. Even gold is suspended in sea water, a tiny percent of an ounce to every million gallons. There are 165 million tons of solids in each cubic mile of the ocean, making it, an expert says, "the world's largest albeit diluted continuous ore body."

Fewer species exist in the ocean than on land: no insects, for example, no highly developed plant forms, but vastly more abundant life. Five-sixths of the planet's total living matter is in the sea, most of it in upper ocean levels, although countless varieties of habitats are available.

The ocean is not just one great big bathtub, as one irreverent scientist puts it, but a colossal, three-dimensional space, varying in depth, salinity and temperature —cold at the bottom where the sun never reaches, warmer at the top. Most species are meticulously adapted to the particular space they occupy, and although there are mechanisms superbly arranged to mix the water—essential for life, as will be seen—each population is likely to stay in its own bailiwick. Interrelationships link the entire ocean into a vast ecosystem that supports inhabitants from Melville's "portentous and mysterious monster that rolled his island bulk in the wild and distant seas" to Cousteau's "rainbow of color . . . the breath-taking splendor of the coral sea which has a profusion of life unmatched by any other area in the world."

All that the first transatlantic sailors knew about this gigantic space was that it was deep, too deep for the two 100-fathom lines spliced together with a deep-sea lead at their end which Columbus ordered overboard thinking he was close to an island. It was a primitive method of sounding, observed by Herodotus in use by the Greeks in 2400 B.C. and still in use in the early 1900s. The coming of electronic instruments and such inventions as the man-made deep-sea fish *Nautilus* in the 1950s and the bottom explorer *Trieste* that went seven miles down in the sixties revolutionized our concept of the depths.

The crucial fraction of the sea rolls across the shelf that borders most continents. Off the Atlantic coast the shelf is quite wide, 50 to 100 miles offshore to a depth of anywhere from 300 to 900 feet, a natural feature which has kept superships away from east coast harbors, their great bulks unable to pass over it. From it, a continental slope descends sharply to the deep ocean floor, now known to be a mountainous place, its crust thinner than the land's and, it has just been discovered, in places newer even than the water it holds. This startling twentieth-century find is added to another—the Mid-Ocean Ridge, a 40,000-mile mountain chain encircling the globe at the bottom of the ocean. Thought to be the greatest single geographic discovery of all time, it excites new theories about the surface of the planet, how it has changed and is still changing.

Like most great discoveries, exposure of the still-developing grandeur of the sea's design has its price. Having turned up more exploitable resources than anyone expected under the waves, it encourages the grab to take what's in the ocean out, and hastens the insistence on putting what's out, in. Major controversies between neighbors, states, nations, merge into what the *New York Times* calls "one overwhelming problem of establishing new regulations," a race against global time before competition for the sea bursts out of control.

At stake is the continuing existence of plankton, the microscopic plants and animals that swarm in every mouthful of sea water, invisible, unknown and unsung until they were found just a century ago by a German biologist trying to net starfish eggs. Their vast quantities make the ocean a living soup. Their function makes life possible.

Plant plankton—99 percent of ocean greenery—provides the basic food for marine life. These tiny green cells, perhaps one-thousandth of an inch in diameter, float where the sun shines, throughout the shallow water of the continental shelf, on the surface of the deep sea. Converting the sun's energy by photosynthesis, they absorb nutrients from the sea. They take in carbon dioxide, produce oxygen; the animal plankton they feed take in oxygen, produce carbon dioxide. Small fish feed on the animal plankton, larger fish on the smaller. The food chain according to size is now thought to be more a web than a chain, owing to frequent feastings across the board; whales, for example, the ocean's largest, have been known to engorge fields of plankton, the ocean's smallest, instead of staying with a diet of larger fish, just as man, the end of the line, enjoys sardines as well as tuna.

For all the billions of years of their existence, plant cells have produced more oxygen than animal plankton can use, a third of the air's oxygen. Without it, we would be unable to breathe. From the beginning, ecologist Eugene Odum tells us, this excess was released into the earth's atmosphere, which until then contained only deadly volcanic gases such as surround certain lifeless, waterless planets today. When enough oxygen had built up (about 3 percent of the present level) in the so-called pre-Cambrian period, an "evolutionary explosion" took place. The single-cell population evolved into new multicellular forms—sponges, corals, worms, shellfish and the rest.

Millions of years thereafter, the plankton had released almost as much oxygen into the air as it contains today, enough for a variety of living forms to fill the sea and invade the land, where, eventually, terrestrial greenery generated additional oxygen and food, spurring the evolution of animals, from dinosaurs to man. Multicelled plants—seaweeds, kelps—attached themselves to the bottom or the rocks of the coasts. Columbus was a few centuries too soon to know that the one he called sargassum is an enlarged plankton, the only plant plankton visible to the naked eye, which turns out to be a common offshore brown seaweed that grows in the warmer regions of the globe and can be found washed ashore by the surf. Too delicate to sink a ship or anything else, it reproduces itself in the almost sterile Sargasso Sea, center of a wheel of currents, by breaking off new shoots to form a permanent floating population of some 7 million tons.

Everything in the ocean decomposes into its elements and is recycled. Over the continental shelf where plankton plants make the sunlit waters a thick soup, they themselves absorb whatever comes into the sea from the land. In the open ocean, where they cannot go down to the sunless bottom, a springtime mixing that brings nutrient-rich water up from the depths, called upwelling, causes a vernal enthusiasm in the plankton that is, if anything, more fevered than our own, as they devour the nutrient feast—remnants of plants, fish skeletons, contributions from passing birds, ships, from rivers that empty into the sea and whatever we choose to put into the water. Miles of sea will take on the green or yellowish color of the minuscule cells as they grow and multiply, sometimes doubling their number overnight during this seasonal bacchanal. Animal plankton rush in to gobble up whatever plants they can find, then eat each other, only to be eaten by the first hungry passing fish. It takes a lot of plant plankton to make a fish; each time one form of life devours another, much of the volume is lost in the process; minerals remain, becoming increasingly concentrated in larger fish.

Some upwellings are caused by winds and currents. A famous example occurs off the coast of Peru, where enormous numbers of anchovies feed on plankton animals that feed on the plankton plants that are enjoying the cold water's nutrients, making it the world's most productive fishery. Another planktonic explosion takes place in Grand Banks waters, shallow enough for the sun to reach the bottom and thus crowded with plankton plants. They are regularly doused with the riches of the chill Labrador Current, where it encounters the warm Gulf Stream, with the expected happy results for plankton, fish and fishermen, a circumstance which took four centuries from recognition to explanation.

Plankton numbers defy comprehension of minds accustomed to an outside limit of six zeros. Investigators say one plankton animal eats 100,000 plant plankton in a single meal, that a herring might consume 60,000 in a gulp; another says one plant plankton gives rise to 100 million descendants in a month, that there are billions of them in every gallon of water, that a blue whale weighing 100 tons will have consumed 10,000 tons of a certain animal plankton called euphausiids, which, in turn, will have eaten a million tons of smaller plankton.

A million tons of plankton? It defeats the imagination to visualize one ton, or even an ounce of cells not much bigger than a pencil point.

All these billions are ingesting what washes into the sea. In just a few years of this twentieth century we have changed their diet. We give them the products of the mass society: human excrement, treated or raw, chemical industrial wastes, oil. Millions upon millions of gallons of this mix pour into offshore waters from cities and rivers every day, or are dumped into harbors and beyond.

Plant plankton do not discriminate between what may please or poison those that eat them. They continue to ingest what's in the water as they have always done, unless photosynthesis, for one reason or another, becomes impossible. Southeast of the Ambrose Light off New York and New Jersey shores, about five million cubic yards of sludge are dumped every year, residue from sewage treatment plants for millions of New Yorkers. The result: in 1973, plant plankton and all the rest of marine life smothered; in 1976, the stinking whiting catch revealed the miles of "dead sea."

If plankton survive, serious considerations arise from their new bill of fare. "Toxic or enzyme-active metals might enter the food chain," a scientist studying the sludge problem says. Other possibilities he lists are hepatitis and encephalitis viruses, coliform bacteria in high enough concentrations to spread meningitis and polio. If the plankton are taking in oil there are other disastrous results. As an oil spill spreads and breaks up into smaller droplets, it may fool people into thinking it has disappeared, but plankton, enjoying small oil particles, know it has not. A 1971 oil spill in Maine's Penobscot Bay killed twelve thousand bushels of clams, more than half of all the clams in Long Bay. They did not simply smother in oil; many with cancerous tumors were discovered in the ill-fated bushels.

We don't yet know the full range of oil's interaction with the sea, Noël Mostert, author of *Supership* says, but what we do know is bad enough. "All crude oils are poisonous to all marine organisms," he says. Refined oils are worse. One hundred sixty thousand or more gallons were spilled off Falmouth, Massachusetts. "Three days after the spill oceanographers trawled the area and found that 95 percent of their catch was dead. A year later," Mostert continues, "life on the seabed was still dying." The shellfish beds remain closed to harvesting today. Effects of oil spills go far beyond the immediate area, beyond what we can recognize as oil, beyond what we are presently able to measure. The most critical aspect of all is the effect on plankton of the soluble parts of oil. These can sink to the bottom, mix with the nutrients, join their poisons to the joyful spring upwelling. "The dissolved hydrocarbons that vanish invisibly into the sea . . . are virulently toxic," *Supership* says. "Can the flora and fauna tolerate the steady invisible toxicity . . . ? We don't know."

Oil from tankers combines with the torrent of wastes from cities to form a supersludge a million square miles out into the Atlantic and down to the Caribbean, a 1974 report says. In New York City's harbor, that "very pleasant place" where Verrazzano anchored, it would cost billions of dollars to clean up the water. Some officials thought the shocking pollution levels which have put many surrounding beaches on the Health Department's restricted list could be alleviated by dumping sludge further offshore. A ruling ordering Du Pont to move the dumping site of a pesticide residue from 16 to 106 miles offshore was disputed by five Du Pont executives, who said that moving the site out "would not help the environment."

They may be right. The branch of the federal government that oversees the seas—the National Oceanic and Atmospheric Administration called by its acronym, NOAA—has a Manned Undersea Science and Technology program with another set of ironic initials, MUST. It studies the ocean bottom in a new two-man submersible vehicle with a transparent nose. It has seen that storm currents move sea-floor sediments, upsetting earlier theories that what is dumped stays where dumped. Currents can carry sludge for many miles, MUST tells NOAA, which tells the world.

Not everyone pays attention to MUST. Harried officials continue to authorize ocean dumping, promise to ban it in the next decade. "There is no place else to put it," one says. Incineration pollutes the air; ground fill pollutes surface water. Where else is there? We pay small attention to the possibilities of recycling, hope instead for a newly invented solution before an irreversible crisis in the sea.

The ocean is in decline. There is little doubt that we have just about used up the generous margin of error it provides. Imposing our products on the crucial offshore waters, we threaten the very existence of those billions of prehistoric plant planktons that first supplied the world with oxygen, that have forever multiplied, furnished food for sea animals, oxygen for water and air. We tamper with a system that once seemed too mysterious and too gigantic to affect. Having comprehended the size and solved many of the mysteries, we now begin to know the danger in disturbing the hair-trigger balance of the awesome sea.

4
WETLANDS

The sea concentrates its riches in marshes, swamps and mud flats. Here, by virtue of its moon-regulated, twice-a-day tides, it fertilizes and cultivates its version of a garden, bringing in nutrients on the rising tide, mixing them with the sediments of the marsh, the mud of the flat, pulling out what the garden has produced as the tide ebbs.

Compared to the vast dilute sea, even to its crucial coastal fraction, wetlands are small, highly organized spaces, efficient beyond belief. They exist wherever the sea reaches in with its tides, at the edges of estuaries where salt water meets fresh—bays, lagoons, brackish ponds—behind sheltering sand bars or barrier islands. Into all these waters, and to the open ocean beyond, tides bring the gardens' produce in great quantities.

These small spaces force us to confront the same confusions that exist in our present use of sand, sea, of the entire coast, but with a difference that makes the confrontation more difficult. To decide what to do with wetlands we must first unlearn an attitude that dug itself into American minds as our forefathers went about settling this country. It is a strong and persistent idea, understandable for its time, intolerable for this time.

Marsh, mudbank, swamp—to us the words have a pejorative impact. "I come from the Everglades," a well-wisher told a recent U.S. president. He shook her hand, walked on, shouting back over his shoulder, "Drain 'em!" a not atypical reaction. It is reinforced by the reputation of wetlands as mosquito-breeding grounds and by the increasing number of marshes that are odiferous in the extreme, the result of the failure of their natural self-cleansing mechanisms due to pollution. A citizen touring a wildlife center with a clutch of children sees a sign "MARSH" and turns back. "Don't bother," she says to the children, "nothing to see but a swamp."

This prejudice began logically enough. As the country became more and more settled, land was thought to be useless if not built upon. It was land wasted, wasteland. People thought that they no longer needed the wetlands for food, putting aside the tradition that had started with the first known occupants of the coast. Archeological finds suggest early peoples at the edge of the marsh, fishing, trapping, hunting, eating shellfish, making ornaments out of bits of purple shell, or white. A certain high bank on the Georgia side of the Savannah River "composed entirely of fossil oystershells" was discovered in the eighteenth century by William Bartram, who, with his father, John, is renowned as a naturalist and explorer. The shells, Bartram noted in his journal, are "internally of the color and consistency of clear, white marble. [They] are of incredible magnitude . . . their hollows sufficient to receive an ordinary man's foot. They appear all to have been opened before the period

of petrifaction . . . are undoubtedly very ancient or perhaps antediluvian."

In Boston recently, workmen excavating for the sub-
way found an enormous and complicated structure built
of poles that turned out to be a fish weir in the ancient
course of the Charles River. Buried in twenty-six feet of
sediment, which had preserved it for five thousand years,
it is remarkable evidence of a large prehistoric colony
living near the estuary and understanding at least some
of its bounty. Later colonists, too, built fish weirs with
the same hungry idea as their predecessors in the neighborhood thousands of years before. "They take a great
store of shad and alewives," an observer notes in 1634.
"In two Tydes they have gotten one hundred thousand
of those fishes: This is no smalle benefit to the plantation." They harvested salt hay from the marsh, otters,
muskrats and mink for trade, fish to eat.

The wetland-wasteland concept that would wipe out
this simple pragmatism was born of expansion and a
burgeoning population. In the mid-nineteenth century
Congress passed Swamp Lands Acts to encourage the
states to "reclaim" wetlands. Since then the country has
lost half of them through draining, filling, dumping old
cars and garbage on them, turning them into something
else. In the 1970s, 300,000 acres per year disappear; the
leading cause of continuing loss, according to the U.S.
Department of Interior, is housing developments.

It took a century to discover that the exploding
coastal population needed wetlands as wetlands more
than ever. In the 1950s, investigators found that they
were not wastelands at all but, indeed, the heartlands of
the estuarine system.

These low-lying sea-washed spaces are now thought
to be the most naturally productive acres on earth. Most
fish, shellfish, shorebirds, some mammals depend on

them for all or part of their lives. Together, wetlands and their adjoining waters make a safe spawning spot, a nursery protected from predators, a lifetime environment, depending on the species. Migrating birds seek them out along the entire Atlantic flyway; in fact marshes have existed for so long that certain species have developed life patterns that depend on finding wetlands all along the coast. Oysters, for example, spend their entire lives in the estuarine system, as does the spotted sea trout. Pink shrimp spawn in offshore waters; almost immediately the tiny larvae begin to move west, sometimes as far as one hundred miles, to reach the safety of the estuary, where they feed so voraciously that they grow from half an inch to full size and then return to the sea as adults. Young salmon migrate to estuaries from high in the rivers where they are spawned; baby bluefish are said to grow an inch a week on food from the marshes.

Using minuscule amounts of radioactive material, banding and other trace methods, recent studies assemble some provocative statistics. Eight of the twelve fish most important to sportsmen and commercial fishermen depend on functioning estuaries; of the 800 million pounds of fishery products landed in New England a year or two ago, 500 million relied directly on tidal marshes.

People need fish. They also need the wetlands' other functions. Spongy wetlands stabilize shores, absorbing flood water, releasing it slowly, thus protecting the land behind them. They are natural water purification systems. A healthy 1000-acre marsh, one investigator says, can purify nitrogen wastes from a town of 20,000 people.

The more people who live on the coast, the more depends on the health of its estuaries. But people now use the productive estuarine heart—wetlands—in ways inimical to continuing function. Florida's large mangrove swamps and salt marshes, for example, produce the state's excellent salt-water fishing, one reason for its vacation and retirement trade, a cycle that brings to the wetlands the dredgers, fillers and builders of lagoon condominiums, high-rise apartments, and other leisure homes, destroying the source of food and shelter for the fish that brought the people to Florida's shores. The circle is vicious; the more people, the more need for functioning marshes, while there is more disruption and destruction of marshes by more people.

There is no way to keep just one part of the estuaries intact. All parts of the system—wetlands, rivers, marshes, ocean water brought in by the tides—are intertwined. What they accomplish in a complex but orderly fashion is the conversion of solar energy into other forms of energy—food—with the help of gravity. Putting aside man's role in this system for the moment, the pure form becomes visible.

Pulled by the earth's gravity, rivers carry minerals and decaying organic matter from the land down to sea level, and ocean water, a 3 percent salt solution with its own load of sediments, is brought in by the tides energized by the moon's gravitational field. Where the two meet in the estuary they mix; the water, loaded with nutrients, becomes brackish, a combination of fresh and salt, and life energies explode. Marsh grass incorporates the sun's radiance and the nutrients brought to it by the tide, stores this energy in its tissues. When blades of the grass die, they are broken down by microscopic bacteria into tiny particles—detritus—nourishing plant plankton so successfully that huge quantities of them float in the waters near the marsh. It is here that this primary link in the aquatic food web is firmly established.

Thus the marsh is a self-contained, self-renewed, food-producing system, self-sustaining because turnover of nutrients is rapid, being constantly circulated through the system by the tide and the life cycles of plant plankton, bacteria and the like, which are relatively short.

From coral reef to oyster reef, mangrove to salt marsh, plankton to pelagic fishery, all forms of life are meticulously adapted to what the estuaries can provide —sun, water that regularly changes temperature and salinity with the tide, nutrients to grow on, and that central elusive catch-all, water quality. Adaptation is remarkably selective; it is found that oysters can live in water temperatures that range from 34 degrees to 90, while striped bass must have their estuary between 55 and 75 degrees. The blue crab puts up with salinity from zero to 34 parts per thousand, the shrimp likes it in the narrow range from 34 to 36.

Best adapted to the rigors of sun and salty water is the marsh grass *Spartina*, cordgrass that whips your legs as you walk through a salt marsh, the only land plant (except mangroves) that can thrive with its roots constantly soaked in brine, and its leaves doused by the tide, in east coast wetlands. Since scientists discovered what *Spartina* does, its every aspect has been the subject of intense examination. It was surveyed by infrared photography from a NASA aircraft around Delaware Bay; it was peered at by marine biologists in a Georgia research center for fifteen years. It is described in some of the extensive literature almost as a person. "*Spartina* does very well in its difficult situation," John and Mildred Teal say; "in spite of the energy it must expend in desalting its water, it still manages to do a lot of growing." A salt-marsh guidebook says one variety "prefers brackish water." Eugene Odum counts stalks: *S. alterniflora* grows forty strands per square meter, its relative *S. patens* two hundred. The former, able to stand wetter conditions, lives best along creeks and grows to ten feet; the latter flourishing on higher ground, once was cut for salt hay, falls over when it ages, forming a characteristic mat.

"Water is pulled into the plant from a membrane on the roots which excludes most of the salt," the Teals say in *Life and Death of the Salt Marsh.* The remaining salt in the sap is secreted by special glands on the leaves; mangroves, also flourishing in salt water, have the same mechanism. Having solved the fresh-water problem it must also provide itself with oxygen, our *Spartina* guides say, which commodity is in short supply in salt-marsh mud, being used up by bacteria at work there. Remarkable *Spartina* pipes oxygen to its roots from openings in its leaves, which it closes at high tide to keep the pipes from flooding. Using tidal energy, it does not have to expend its own, as humans do, on cycling minerals and transporting food and wastes within its system. "It functions with great economy," the Teals say, "a success story of a most complicated nature."

The true success of the salt marsh is what it provides to its estuary customers. "Why do such highly prized sportsfish as striped bass and bluefish congregate at the mouths of tidal creeks and inlets? How do the bays and estuaries support such a large population of edible shellfish . . . ? Why do the young of flounder and menhaden abound in the tide-marsh coves and salt ponds?" a biologist queries. The answer is obvious to *Spartina* watchers. Its fundamental product, detritus, nurtures great colonies of plant plankton. Mixed with algae and bacteria, detritus also feeds all kinds of predators—insects, flies, fiddler crabs, snails—and filter feeders such as oysters, mussels and clams. Largest detritus eaters are small fish, mullet and menhaden used by humans for fertilizer. Mammals and birds eat the detritus eaters. They, in turn, are eaten by even larger estuary hunters —ospreys and other hawks, big fish, people.

Production of detritus is prodigious because *Spartina*

growth is. It builds a complex root system which eventually decomposes into peat, new roots forming on top of the old layers. The marshes move by virtue of this root system, as it traps sediment along the shore from the rivers and the sea, keeping pace with the movement of the shore itself.

These marshes are an intriguing index of coastal maturity; Maine's 15,000 acres are but a fraction of Delaware's 115,000 acres, which are less than a third of Georgia's 400,000 acres. The reasons for this difference lead us back to how the coast was formed. In the north, as far down as Cape Cod, drowned river valleys are in an early stage, a steep, sculptured shoreline awaiting the deposit of additional sediment, recovering still from the harsh glacial ice that swept northern slopes clean to deposit its sediments as far out as Georges Bank. From Cape Cod to Cape Hatteras marshes are more developed, and protective spits and barrier islands are forming, while from Cape Hatteras south the low coastal plain, brushed by the Gulf Stream, ends in barrier islands. Here, a large amount of sediment is deposited by rivers, evolving all the way to coastal marshes.

It is possible now for the first time ever to read the marshes and know how and why they form as they do. Tracing the case history of Barnstable marsh on the north shore of Cape Cod, its underpinnings laid down by *Spartina* roots some 4000 years ago, a Woods Hole scientist reconstructs the marsh as it grew during eras he identifies as those of the Trojan War, Alexander, the Crusades, the present. (The young marshes further north go back only 2300 years.) As sand and sediment accumulated, the marsh gradually filled in the harbor, adding layer after layer of peat, which is deepest where the marsh began when the sea was some 18 feet lower than now, thinnest at its newly forming edges.

Largest in Maine is the celebrated 2700-acre Scarborough marsh; most are smaller, ten-, twenty-, or fifty-acre growing areas with down-east names—North Goosefare Bay, Northwest of Buttermilk Cove, West of Bailey's Mistake—dotting the rugged coastline. Southern marshes, without a glacial history, are older, deeper and enormously more extensive. Of the thousands of miles of marsh that once covered the coast from New York south, most in or near our cities have long since been transformed to garbage-based dry land. But parts of the green ribbon survive, widening behind protective barrier islands to the magnificently productive Georgia marshes. "How ample, the marsh and the sea and the sky!" exclaims Lanier's "The Marshes of Glynn." Still mostly intact, marshes in Glynn County and elsewhere, with their tall stands of *Spartina,* green in summer, yellowed in winter, the golden mud banks exposed when the tide ebbs, stretch from two to ten miles between the mainland and the barrier islands.

In the 1950s, the remarkable size and sweep of Georgia's marshes and what is often described as their mystery, much of their business being conducted under cover of the close-growing *Spartina,* attracted the attention of Eugene Odum and the University of Georgia, where he was a professor. "The notion came to us, in those early days," Odum recalls, "that we were in the arteries of a remarkable energy-absorbing system whose heart was the pumping action of the tides. . . . Does nature routinely exploit tidal powers as men have dreamed of doing for centuries?"

Odum and his research team made a now-famous calculation. Georgia estuaries, by their figures, produce the equivalent of ten tons of dry organic matter per acre per year. By comparison, the average wheatfield produces one and a half tons. The estuaries manufacture twenty times as much as a comparable area of the open ocean, five times as much as coastal waters. A decade later his surmise was established, confirmed by other investigators as was his quantification of marsh production—the much-advertised ten-ton-per-acre-per-year figure.

The ensuing excitement about wetlands spread quickly from the scientific to the conservation-preservation communities, and in the generalized terror that we were losing even more than oysters, flounder, shorebirds, muskrats *et al.*, it began to be popular to endorse the existence of wetlands, to protect them. Wetland literature flourished, magazines published and republished the ten-ton and other figures, there were eloquent pleas for mercy for the marshes.

> *Ye marshes, how candid and simple and nothing-withholding and free*
> *Ye publish yourself to the sky and offer yourselves to the sea!*

the romantic poet Sydney Lanier wrote in a worshipful peaen to what he called "that vast sweet visage of space."

We knew enough and were scared enough to try to protect wetlands, to reverse the ingrained "Drain 'em" reaction. First steps were taken; in 1963 Massachusetts passed a coastal wetlands protection law, followed by most New England states, later by some southern states. Federal agencies announced their good intentions toward wetlands; the Department of Interior made a national estuarine pollution study in 1969, the Department of Commerce included wetlands in its coastal protection program, the Environmental Protection Agency made it a policy to minimize alterations in these besieged spaces. In 1972 the Federal Water Pollution Control Act, now under attack, gave the Army Corps of Engineers responsibility for protecting the nation's rivers and wetlands, a job that the construction-minded Corps has so far allowed to languish.

There have been some salubrious effects. Serious efforts, the first in history, are under way to build new wetlands. The University of Maryland tried to produce five acres of salt marsh in five years. Rutgers Marine Science Center used silt and clay from dredged portions of the Intercoastal Waterway to fill shallow waters next to marshes, transplanting marsh grass. The reluctant Corps of Engineers sponsored the first man-made salt marsh in Chesapeake Bay in 1972, planting some sixty thousand *Spartina* seedlings by hand on a newly created sand flat. If this use of dredge material from harbors and rivers succeeds on a large scale, it can make a difference on the wetlands map in the twenty-first century.

Despite these efforts, laws, statements, and studies, wetlands continue to be lost, dredged, filled in, developed much faster than new ones can be made. Estuarine health continues to decline, pollution levels rise. If sea water is prevented from flooding a marsh by a causeway or other obstruction, an irreversible acid reaction sets in and the marsh will not go back to normal even if flooding is restored. Dredging stirs up bottom sediments that rapidly remove oxygen from the water with devastating effects on fish. Wetland supporters are surprised and disappointed by the response to their discoveries. It is hard for them to believe that the ten-ton figure, the proven facts of wetland function, fail to arouse others, that, except for a few gestures and a considerable tangle of added red tape, traffic in these lands continues almost as before.

We have not overcome the past century's prejudice against wetlands and the later appetite for shore front. In the late sixties, North Carolina was losing an acre of wetlands a day, Maryland a thousand acres a year, Connecticut 15 percent of her remaining wetlands in a decade. Conflict of use became more and more evident. Where Indians affected wetlands not much more than a marauding bear or grazing deer, almost everything we do in or near our estuaries degrades them. By the early seventies, the predicted losses had occurred. There were fewer mosquitoes but also a scarcity of certain species of fish, shellfish and birds, of usable sheltered water, of places to swim, fish, hunt on the coast. In their place are multiplying port facilities, industries which find estuaries convenient locations, more and more dams, developments, man-made environments.

Wetland ownership makes the situation more difficult. Salt marshes and bay shores are in a different circumstance from free-running sand and sea. People can own wetlands and do. Most estuarine land on the east coast is in private hands. Although the function of each wetland acre affects the entire population, its fate relies on the will of its owner. "The owner of a salt marsh becomes a public servant," a conservation tract says. "The richness of his marsh flows into the sea—and is harvested by the public." In 1977 one economist suggested that the wetland owners be reimbursed by the public for maintaining the natural state of these lands because they are used as a home by the public's fish, thus providing a collective good. Without some such plan, not many choose to keep these lands intact; in today's economic crunch, few can afford the altruism to protect for the public good products like detritus that cannot be seen, harvested, sold or eaten.

Pristine pockets remain. Muskrat pelts are nailed flat to a barn door on an 1800-acre farm, mostly marsh, that spreads along the west side of Delaware Bay. The present occupant, ninth generation in his family to live on this land, says the boys get the muskrats after school, 600

or more during the winter. "A good black one brings
$4," he says, "a brown one $2.50." Muskrats, he says, are
a remarkable resource: "Trap 'em and they come back
better." This farmer's respect for the marsh is enduring.
Muskrats from the marsh saved him during the depression, bringing in a cash crop of $1500 a year. This, with
money from hunters who pay at least $25 a day to shoot
plentiful ducks and geese from his blinds, and from a
harvest of terrapin in the streams, oysters in the pond,
makes a living even in the hardest times. Most of his work
consists of keeping the wetlands in their natural condition, occasionally impounding a pond or clearing a road
through high land. Where his grandfather burned the
marsh vegetation, thinking it good for the muskrats, he
plants trees.

Bordering the old farm are wetlands newly acquired
by the Shell Oil Company for a refinery not yet built. The
neighborhood fought hard to keep Shell out, eventually
lost. "Shell didn't have a snowball's chance till the energy crisis came along," the farmer says. He enjoys a
one-man revenge; the wall of his hunters' shelter is pa-
pered with pink rectangles which on closer inspection
reveal themselves to be uncashed dividend checks from
the one share of the Shell Company he bought. "At
least," he says, grinning, "I can worry their bookkeepers
some."

We begin to destroy some of the things we like the
best. "The Atlantic salmon has almost completely disappeared from the east coast," the Department of Interior
states, "a classic example of the damage inflicted on
fisheries by [estuarine] biophysical modification." As for
the oyster industry, it is "a continuing story of depletion
in absolute quantity and decline in the usefulness of
remaining beds in nearly all estuary areas that naturally
supported oyster populations."

The succulent oysters of Delaware Bay, for example,

have been savored by people for three hundred years at least. An early traveler to Lewes, a town on the west shore of the bay, notes that he had "some of the largest oysters and cockles I ever saw in my life; some of the former were six inches diameter out of the Shell, and very well tasted." Their shells were used for roads and for lime in the soil, and oyster gathering became a profitable business. Mindful of their estuarine treasure, bay counties protected oysters with ordinances in the eighteenth and nineteenth centuries, establishing a closed season, prohibiting dumping of shells in creeks, making dredging illegal. The oyster industry thrived so well that by 1860 it is reported that on the 4 P.M. freight to Philadelphia two locomotives were needed each day, and eight freight cars were filled with oysters.

The word spread. Oyster pirates from New England came to the bay in vessels armed with cannon or under cover of night to scoop the delectable shellfish from their beds. Oyster wars raged. There were fighting and bloodshed over the tasty bivalves until finally, an observer noted at the time, out-of-state trespassers were greeted by a steamer cruising the bay and coves bearing the inscription, "Thus Far Shalt Thou Come and No Farther." When oyster culture developed, the state planted beds to add to the produce of the naturally growing oysters that found Delaware Bay a particularly nourishing environment.

Oysters lie on the bottom of estuaries, fixed to the shells of their ancestors or other hard surfaces, waiting for the tide to bring them nourishment from the marsh. They are filter feeders, straining particles of food from the water by their gills, concentrating nutrients and minerals in their digestive systems. They also concentrate whatever poisons are in the water that flows over them from the marshes, are quick to show the effects of a change in their diet. (In Japan an investigator found "green oysters," with ten times as much copper as oysters from uncontaminated waters, three times as much zinc.)

Delaware oysters have been attacked by MSX, a protozoan parasite for which there is no known treatment. And oyster drills, snails with a predilection for oysters, became a problem as more dams were built in more rivers, allowing the drills to migrate upstream with the salt water and attack oyster beds. While concerned officials search for disease-resistant oysters to plant in the estuary, oystering as it was in Delaware Bay is over. In 1973, residents of a small bay town, who had always made a good living in oyster-shucking houses, went on relief.

For a small state, Delaware has many distinctions in addition to its once-plentiful tasty oysters—the largest percentage of wetlands of any state, the greatest concentration of oil refineries on the east coast, the second-largest port in commercial tonnage, or so its governor's task force on the coastal zone reports. It also has a notable population of mosquitoes (fifty kinds) and biting flies (seventy kinds) that breed in its tidal marshes, making the bay shore "a veritable hell to man and other animals" from June to October, the task force says. Developers press the state to get rid of these pests; in response it ditches and drains many marsh breeding grounds. Some marshes are bought by oil companies that don't seem to mind insect or other antagonists—the thousand acres acquired by Shell, another thousand filled in by Tidewater Oil, three hundred bought by Sinclair for a tank farm and unloading port, many more.

No part of Delaware is more than eight miles from the wetlands, so what happens there is important to all the state. "How many acres of these marshes can we afford to give up without affecting the rich coastal environment?" Franklin Daiber, professor of Marine Biology at the University of Delaware, asked in the early 1970s. "What role do these marshes play and what effect does man's manipulation have?" The answer is still not fully known but meanwhile conflict of use in the Bay affects more than oysters. Terrapin, fish, migrating ducks and geese, muskrats in the marshes diminish as industrial activity multiplies, despite a landmark law passed in 1971 barring any new heavy industry from these shores. Huge oil refineries tower on the banks like gigantic Rube Goldberg contraptions. The "Chemical Capital of the World" hastens disruption of the quality of life which men who make their fortunes from its products have come to expect.

With the bull market in waterfront land skyrocketing and the population of coastal counties soaring, cold cash for wetland acres has its appeal—more appeal, it appears than public service. It is a problem that magnifies the dilemma facing other elements of the coast. But salt marshes and their nearby shallow coves and creeks, teeming with eager, marvelously ingenious forms of life, are poignant persuaders. Gradually, perhaps too slowly but with persistence, they insist that we comprehend what we are losing.

5
BIRDS, FISH, AND OTHER COASTAL INHABITANTS

The coast, a London gentleman observed, is "a damp sort of place where all sorts of birds fly about uncooked." He spoke for generations of hunters (later called sportsmen). But when too many birds had been shot for cooking or other purposes, and, in the late nineteenth century, were on their way to extinction, the sparkling graceful terns and slow-flying dusky herons roused man's tendency, as Darwin put it, to feel tenderness for lower orders, and the shooting stopped.

In the 1970s, the birds and other coastal creatures are again menaced. The threat is far more deadly than the hunter's gun. Their world is under attack.

The world of sand, sea and wetlands is the only world for coastal birds, for fish, for whales, seals and other marine mammals, for the masses of invertebrates from microscopic plankton to lobsters and clams. Unlike man, who can live almost anywhere on earth—mountain, desert, forest, or the fortieth floor of a concrete tower—

coast creatures cannot move from one environment to another at will, or at all. They are the products of a long evolutionary process that has precisely calibrated them for a naturally functioning coast.

They exist in marvelous variety. Life reaches into every ecological niche, fitting a species to each of the myriad coastal circumstances.

Consider the decorator crab that covers its body with bits of seaweed which obligingly continue to grow, effectively concealing the crab as it stands on its long legs and ·"keeps up an almost continuous waving back and forth," a biologist says, "to look like the surrounding seaweed waving in the current." Or the striped bass that spawns 15 to 30 miles upstream in 62- to 65-degree water; its eggs will hatch only if suspended for their first 48 hours at this temperature, perish if they touch bottom. Or the ruddy turnstone, a shorebird that has its niche on a stony beach where it feeds on the small animals hiding under the stones it turns with its stout beak. Or the snowy egret, only bird in the pond that shuffles its feet in the mud to stir up its particular food supply. Some birds live on beaches, harvesting the edge of the tide or diving in the waves; some wade in ponds, just as some fish live in shallows, some in the depths.

Birds and fish are the most valued and familiar of the many life forms that have been nature-selected for the shore. For our purposes these two groups can represent all their coastal companions: the burrowing ghost shrimp, the harbor seal, the American lobster living only in a narrow strip off New England, oysters, sea scallops, even the minute reef-building coral. Details differ for each life form but the tumultuous circumstance of their life as it unfolds in the late twentieth century is more or less the same for all.

Remarkably determined travelers, fish and birds

move up and down the coast to find satisfactory living conditions the year round. Some birds migrate from the Arctic to the Argentine pampas, some fish from high in fresh-water rivers to the salty depths of the Outer Continental Shelf. They have accustomed stopping places on these journeys. Among the 146 duck species, for example, canvasbacks drop down to the bay near La Guardia Airport, Roger Tory Peterson says, while widgeon are grazers, stopping at certain golf courses on their route; brant alight on salt bays where eel grass, their preferred food, grows, while scoters dive for mussels on sheltered beaches. Highly specialized, they move in flocks, nest in colonies, depend on the food supply generated by the shore and on a particularly delicate balance of numbers to perpetuate themselves.

Fish move all through their lives. Seventy percent of the fish we eat start off in bays, coves and inlets; after months or years, they search out the freedom of offshore waters, the crucial ocean fraction. They have evolved to thrive in the ocean, "by far the easiest place in the world to live," a marine scientist says, with its even temperature, ever-present food and minerals, and stable chemical composition. Marine life depends on this constancy, virtually unchanged over eons.

The species of our time are triumphant survivors, whether by intrinsic accident or extrinsic pressure, of many thousands of varieties that have come and gone through the millennia. Their self-perpetuation appears guaranteed. A large oyster can produce 500 million eggs in one season; a 21-pound cod on dissection had some 6 million eggs; flounders average a million eggs each. Natural dangers to eggs floating unprotected reduce the surplus but a plentiful margin survive to produce a new generation. It is estimated that only one cod in a million matures. Even so, the cod population is vast, thanks to the precise balance with its marine environment of the fish that fed the first explorers and named Cape Cod. Along with other species, a Massachusetts colonist notes in 1630, they were "in abundance beyond believing."

Now, suddenly, the abundance diminishes. Mere survival is the question. Will birds and fish survive in the conditions we are creating or will they die out? Research is needed in a dozen disciplines for us to understand the interwoven effects of oil in water, dams in rivers, sewage in estuaries, traffic on beaches and other new circumstances, but there is already some pertinent information about what is happening to coastal creatures and about their capacity to take it.

One stupendous development is that the world fish catch has tripled in twenty-five years, part of the effort to feed the world's incessantly multiplying population. Most fish live in temperate zones over the shelf that borders continents, one-fifth of them in still-teeming numbers off U.S. shores. In the last fifteen years these numbers have been depleted as never before, not by the U.S. commercial fleet of 130,000 fishermen in relatively small, old-time, individually owned boats, nor even by the swelling numbers (up from 4 million to 10 million) who take advantage of the leisure bonanza to go fishing. Although our nation has doubled its consumption of fish since 1950, almost 60 percent of the fish we eat is imported. The hugely increased catch off our shores is overwhelmingly the work of large, technically advanced fleets from China, Japan, the USSR and other nations, roaming the world's oceans in search of food and mining the abundance of the east coast, U.S.A. In the last decade, the catch within 200 miles of the U.S. coast that went into foreign freezers soared from 1.5 billion pounds to almost 8 billion.

The dominant aim has been to catch as much as possible wherever the fish are. Long-range consequences have been hardly known, or ignored. Salmon feeding grounds off Greenland, for example, when discovered by the Danes in the 1960s, soon lured a half-dozen nations; the catch zoomed from 132,000 pounds to 6 million, a near-fatal blow to Atlantic salmon that now return in only decimated numbers to whatever east coast rivers remain navigable for them (of the 37 in Maine, 8 now have salmon). Haddock off New England were taken in massive amounts by new electronic methods which "vacuum" the sea, reducing the stock so severely that it is unlikely to recover. Subsequently the catch decreased from 134 million pounds in 1966 to 2.5 million pounds in 1972. In the mid-seventies, twenty other species—fluke, halibut, striped bass, lobsters, oysters—have been severely depleted.

The race for fish is so furious that in 1976 the nation claimed all fish within 200 miles of its shores, hoping thereby to eliminate foreign competition, even though many fish are, like salmon, world citizens, commuting to faraway waters. To be rid of another competitor, there has been a straight-faced proposal to rescind the 1972 federal law protecting marine mammals, particularly in respect to Maine's seals, which are thought by some to be getting more than their fair share of the sea's bounty. As things now stand, there is little doubt who would win in Man *vs.* Seals.

The world fish catch warns of a fishless future. For the first time the *rate* of increase is declining; zero growth is forecast. How much responsibility attaches to our alteration of fish habitats, how much to overkilling is not known, but the total effect is clear. A committee reports to the President and Congress: ". . . there exists the possibility to fish to extinction."

It is not a new experience. Once before there was

overkilling on the coast, a decline in numbers and species, the endangering of some, the extinction of others. Then the threat was to the wild birds flying about uncooked. True the scale was small but the motive for killing birds was just as acceptable as commercial and sport fishing is today. The legacy of belief in inexhaustible coastal resources at man's command prevailed then as now.

Jacques Cartier, exploring in the sixteenth century, saw islands crowded with nesting birds. In half an hour his crew killed enough flightless great auks to fill two boats, and he thought all the ships of France could do the same with no noticeable effect. He named three small islands after the gannets, goose-size white ocean birds "as thick ashore as a meadow with grass." Two centuries later, James Audubon too was astonished at the gannet-covered rocks, "the nests two feet apart in such regular order that you may look through the lines as those of a planted patch of cabbage." At his first shot, the birds "fell into the sea by the hundreds," Audubon says. Fishermen armed with clubs killed them for bait "until fatigued or satisfied. 540 have thus been murdered in one hour by six men."

Visiting the Florida Keys, Audubon was again astonished by the multitude of birds, "great godwits, rose-colored curlews, purple herons, white ibis," and he noted that his first volley into a flock procured enough food for his party for two days. At breakfast they consumed a heap of ibis eggs, then they waited for the tide to come in across the mud flats, driving the birds before it. "Each of us, provided with a gun, posted himself behind a bush . . . and the work of destruction commenced," Audubon says. The result—a pile of carcasses "resembling a small haycock."

There was no known reason not to hunt coastal birds,

no knowledge of the numbers required to perpetuate the species or of the birds' contribution to the ecosystem (except for the guano birds living in huge colonies off the Peruvian coast, which produced millions of dollars' worth of the world's finest natural fertilizer and were known and protected by the Incas a thousand years back). Dense clouds of migrants winging up and down the coast, like the crowds of fish in the sea, encouraged the belief in an unlimited abundance.

The effects of mass killing went more or less unnoticed until extermination started in earnest. One hunter shot more than 4000 lesser yellowlegs in one season, another, 335 ducks in one day. Millions of sandpipers, plovers, curlews were trapped, or shot, or killed by firelighting, a custom of blinding flocks at night so that hunters could wring the necks of the mesmerized birds. The Parker House in Boston paid fifty cents a bird for yellowlegs, ten cents for knots. By the 1880s there was an additional reason to kill shorebirds, a long-established right. When a certain Richard Hartshorne, a Quaker, bought Sandy Hook, New Jersey, from the Indians in 1678, his rights to the land included "plumaging." Two centuries later, plumes were a new fashionable necessity; the most desirable, after ostrich feathers, were aigrettes, the trailing plumes of the snowy egret, common egret and the roseate spoonbill, grown only at breeding season.

Colonies of these plume producers were left dead among their nests, their trailing feathers snatched from their carcasses. Other shorebirds, too, decorated Gay Nineties heads and hats. A pair of glistening tern wings brought fifteen cents, a skin forty cents. Some 5 million American birds were slaughtered each year for this purpose.

Eventually the alarm sounded. Audubon Societies

founded at the turn of the century urged protection of the birds. The Migratory Bird Act of 1913 stopped the shooting of most coastal birds, limited the killing of waterfowl. It was too late for the great auk, whose ancestors Cartier's crew clubbed to death on crowded northern islands; the Eskimo curlew edged close to extinction; the golden plover never quite recovered.

Many species could and did rebuild their populations, still gamely trekking north to breed, south to winter, scouring the shore for food, fishing in coastal waters. But the proportions of today's unprecedented threat—fast alteration of their environment—loom exceedingly large when set against the birds' multi-million-year history.

The coast birds' attributes developed over many millennia in response to the ancient shore. Evolving, like all birds, from the reptiles, their bodies specialized to suit an exceptionally athletic life during which they are constantly flying, swimming and diving. Powers of touch and smell, hardly needed, are rudimentary, eyesight is better than ours, providing both monocular and binocular vision.

When the general form became place specific, nonflying penguins' wings modified into flippers, while master migrators like the arctic tern developed great flight muscles, colored to "dark meat" by a generous blood supply. Diving birds achieved the capacity to keep breathing while grasping their prey under water: pelicans, for example, have inflatable air sacs, loons can hold their breath for five minutes, dive down two hundred feet. Terns' slim shape and powerful wings are designed for hovering over the water, making shallow, headfirst dives, and each tern subspecies is equipped to pursue its own life style. Roseate terns, according to naturalist John Hay, are built to dive a foot deeper than the common tern, reaching different prey. Sooty tern chicks have evolved to go without food for several days while their parents fish two to three hundred miles at sea, and *their* catch is not available to the noddy terns, which fly only fifty miles out, having to get back to the nest by nightfall. Some tern species nest within reach of the tide, patiently replacing eggs that wash away, some in colonies on a particular rock or sand spit to which the same birds return year after year, a habit known as *ortstreues* (place faithfulness).

Place loyalty can shift. In North Carolina, coastal birds, evicted from their shore nesting grounds by invasions of vacationers and beach buggies, are momentarily, if accidentally, rescued by an unexpected savior, the U.S. Army Corps of Engineers. The Corps dredged the Intercoastal Waterway, piling spoil that established sandy islands, one after another. Now in all stages of development, the islands provide places for many species to nest, from oystercatcher and royal tern, which choose the bare beach of the newest dredge, to hundreds of pairs of herons, egrets and ibis in the dense elder and myrtle thickets of older islands. Eighty percent of North Carolina birds are now faithful to the Corps's dumps, unaware that with a change in dredge policy these new homes too could be gone tomorrow.

Such species, adapting over eons to imperceptible changes in sea and shore, survive for thousands or even millions of years. They may eventually die out. The extinction rate for all wild creatures used to be two species per century. In the last hundred years we have raised that rate between four and fifteen times (the experts disagree) and even these have been small, widely spaced events, so that we do not much miss the now extinct passenger pigeon or the Carolina paroquet that John Bartram noted in 1765 in majestic cypress trees, "commonly seen hovering and fluttering in their tops."

The most specialized species are the most vulnerable to changes in their environment, according to Roger Peterson. At present less specialized birds—crows, jays, sparrows, finches—are thriving in cities and suburbs, adaptable and capable of further evolution, "and that," Peterson says, "is what counts."

By this dictum the future is not cheerful for the highly specialized birds that depend on the coast. Backs to the ocean, they have no place to go if the shore fails them, cannot depend on such chance rescues as the Corps of Engineers may happen to provide. Even migration offers a limited choice; the shore is a limited space. Coast birds travel greater distances than any other bird group to winter in the tropics, breed in the Arctic. The golden plover spends a brief summer up north, then takes off for its South American haven some 8000 miles away; the Arctic tern's record trip—12,000 miles twice a year— may take it down the west coast of Europe and Africa or around the Pole. Like brant ducks searching out eel grass, each species expects to find a familiar source of food en route. But now, spreading megalopolis and other human activity from Maine to Florida and on most of the world's coasts is closing out this indispensable requirement.

The results of overkill, followed by drastic change in environment, challenge the offhand notion that, if we make a few adjustments, the birds will survive somehow and we can safely continue with business as usual on the coast. The situation of fish redoubles the challenge.

Fish were the earliest vertebrates on the evolutionary tree, their ancestors jellyfish (first animals with mouth and stomach) and worms (first with nervous system and brain), found in fossil imprints 700 million years old. By 400 million years ago, fish dominated life on the globe, surpassing invertebrates that lived in the water, attached, crawling or floating, passively awaiting food.

Best-known fossil fish because of its perfectly preserved remains is the small osteostraci. Its flat head and body are covered with bony plates, this heavy armor making it clumsy and probably sluggish, living on the bottom. As in other elaborately armored groups, when competition in the water increased, these died out in favor of less protected and more active forms.

Those fish that gave up armor to move through the water to feed and protect themselves had to contend with an environment eight hundred times denser than air. "The fine form of a typically swift-swimming fish such as the mackerel is admirably suited to cleaving the water," zoologist J. R. Norman's *A History of Fishes* says. Like the submarine, it uses the least possible energy because its cigar shape, bulletlike head, small scales and closely-fitting gill covers make a smooth surface that offers little resistance in its forward motion. When speed isn't important, the shape of the fish reflects that too. The blue shark, which chases its prey, is perfectly streamlined and as blue as the open ocean it roams, while the carpet shark, which lies in wait on the sea floor for prey to come its way, is stout, flattened, with a massive head. "The loss of swimming power," the zoologists say, "is compensated for by the remarkable manner in which the shark resembles its surroundings, its appearance being that of a weed-covered rock."

All marine life is dependent on the everlastingly changeless (until now) composition of ocean water. Senses are attuned to it. In fish, taste and smell—the chemotropic—which are relatively rudimentary in people, are highly developed. Many species rely on chemoreception for locating food and sex partners, for signals that lead them to spawning grounds, others that warn them of danger. Because they have no need to see great distances, they are near-sighted; hearing is their least-developed sense. And because the ocean's temperature, like its composition, is constant, hardly changing with the seasons, fish require no temperature-regulating mechanism, have developed none, derive their temperature and metabolic rate from the water.

Just how constant the ocean has been, at least in the depths off South Africa, was dramatically demonstrated in 1938 when a fisherman, hauling from 40 fathoms, landed a five-foot coelacanth, a fish thought to have been extinct for 70 million years. It differed hardly at all from its fossilized ancestors that lived 345 million years ago. In 1972, another live specimen was captured, intriguing scientists by its very simple heart, large ova (3-½ inches across) and the fact that this very ancient species swam sideways.

During the first days of coelacanth, the ocean as we know it was more or less established, fish were specialized for function and specific habitats and there was little pressure for large-scale adaptation. The last great upheaval was the glacier that sculpted the North American coast and lowered temperatures temporarily, but at a pace that allowed thousands of generations to accomplish the necessary alterations of a gene. Today, a far more violent intrusion—the introduction of PCBs, a hydrocarbon discovered in high levels in striped bass spawned in New York's Hudson River, and of petroleum hydrocarbons, oil, into the biosphere—calls for far faster adaptation—faster than may be possible.

Looking back through the millennia of remarkably slow specialization—from the armed to the streamlined fish, from the primitive winged reptile to the acutely specialized tern—it is no surprise that the coast's wild inhabitants cannot match the pace of man's rapid reconstruction of sand and sea. Some have already departed this world forever; some could react to the environmental breakdown by reversing evolution. The extreme importance of speciation was recognized by Darwin, who was vastly impressed with the "beautiful adaptation" of certain finches, tortoises, armadillos. A world without the natural barriers that have kept such species specialized through the ages—a concrete shore instead of free-flowing sand and dunes, hot, oil-heavy polluted water instead of the historic solution at its constant temperature—is projected by A. J. Cain, British zoologist, as "an appalling welter of hybrids. No single individual would be adapted to any one mode of life." Eventually there would be only one, general-purpose species, Cain says, able to maintain itself but not adapted to do any particular thing with special efficiency. Would it find its way to estuaries for food and nurture, manufacture millions of eggs to perpetuate its species, dive deeper than its close relation, provide us with the assurance of vitality in variety? No one knows.

Not many people would deliberately send the coast's wildlife back down the evolutionary tree to find out, nor would many opt for its extinction, but on the crowded, altered east coast, such possibilities are imminent as the birds and fish demonstrate. Either they go or we change.

We are adaptable, equipped to change fast, governed by different rules from those that order other living things. Man has started off a new kind of evolution, George Gaylord Simpson says in *The Meaning of Evolution*. The old organic evolution that gave rise to *Homo sapiens* had no purpose or plan, no single trend toward higher things, Simpson and others believe. But humans can control their further evolution through knowledge, a sense of values, the possibility of choice. They control the destiny of other forms of life as well, and cause the extinction of other organisms at will.

Our own immediate and pressing needs have determined what the coast would be without much thought for the consequences to coastal creatures and, long range, to ourselves. When the effect of our killing birds could no longer be ignored, we did have the instinct and intelligence to change our sporting ways. More complex reasons for the present disastrous condition require vastly more demanding sacrifices, if it becomes our intent to preserve life on the coast.

The value we put on the coast is best seen by what we have done to transform it and what we plan to do next. The record of our action allows some hopes that there may still be flocks of birds flying low over the shore in the twenty-first century, some hopes for seafood in our diet. But not many.

6
LITMUS FOR THE COAST

The first time the United States ever counted its islands was in 1970. Considering the circumstances of the east coast at that moment, the inventory was an ardently wishful act. Accumulated effects of the modern use of the coast were exploding; evidence of the disturbed coastal ecosystem was blatant. "Say it isn't sludge!" was soon to resound along the beaches. Perhaps the islands, remote, romantic, would provide reassurance that the natural coast still functioned and could be maintained.

The message from the islands was in concentrated form that of the new man-made shore. What happens on the main permeates its islands. They are a reliable litmus for the coast.

The islands are not homogeneous but differ according to ancient origin, as the coast itself differs. Rocky, spruce-covered Maine islands brightened by silvered birch are the tops of granite mountains drowned by the rising seas, not at all like the flat, sandy barrier islands that parallel the southern coast, or Martha's Vineyard, with its boulder-studded hills that mark the end of the glacier, or the remarkably unshakable rock under Manhattan. Such particulars, along with tides, currents, sand drift, form individual island character, the raw material to which man fell heir.

One unique group of islands that started out thousands of years ago with the same geological base and developed the same raw wilderness are now startling in their differences from one another. They exhibit the process of alteration from the primitive state, hold up the mirror to each act of the past half century that changed character and use, lay it straight out why man did what he did—a causeway, a groin, a development, a tax—and the results. This remarkable etiology has a powerful impact.

It exists on the eight barrier islands at the innermost sector of the arc curving in from Cape Hatteras for a sweep around to Florida's tip, the Golden Isles as they are called, Georgia's coastline, part of the string of 281 barrier islands along the Atlantic and the Gulf. More wild coast exists on the Golden Isles in the mid-1970s than in most places. It is still possible to see a Georgia island much as it emerged from the last stages of the Ice Age, shaped by sand drift and rising seas. In subtropical inland forests, agile small island deer bound among the magnolias, red cedars and huge live oaks curtained with gray-green Spanish moss. Wide white beaches border the forest on the ocean side, while toward the mainland sediment collected over centuries has built Georgia's renowned marsh, two to five miles wide, some 400,000 acres of it, said to be the most fertile anywhere. Marsh, forest and gleaming beach rustle with teeming life. The whole has a compelling serenity.

The islands' size (the largest, Cumberland, is a third

bigger than Manhattan) and particular appeal made them specially appropriate land grants from the British sovereign to planters, their first owners, pleased to have large tracts to cultivate. The same attributes found favor with the next wave of landowners and has kept them singly held and singly used for several centuries, the wilderness virtually untouched. One man, one island, was a stable equation.

Today it hardly exists. Some islands have been developed, their character homogenized with the mainland. Others exemplify the currently popular notion of sheltering such pristine pockets under an institutional wing, an undertaking that has coast-specific difficulties, as the islands demonstrate. One is becoming a National Seashore; one teeters on the edge of change. The virgin state on the Golden Isles—as on the Atlantic seaboard—is about to be everywhere undone.

There was good hunting in the marsh for the Creek Indians who roamed these islands as did their prehistoric predecessors—hunters, gatherers, consumers of shellfish—as early as 3800 B.C. Island-hopping Spanish friars converted the Creeks if they could; French traders bargained with them for wild turkeys that they sent home to their monarch before the Pilgrims ever sighted the Thanksgiving bird. Use was decided by natural resource, and this condition continued even after British General Oglethorpe, founder of Georgia, claimed the islands for the Crown in 1733. Soon they belonged to the planters, who raised now-famed sea island cotton and indigo, lived in luxury afforded them by slaves. Only one island, St. Simons, was never singly held. Here Oglethorpe settled certain obligations by a number of grants, unknowingly establishing the foothold for development, twentieth-century style.

The opulent plantation era used but did not impair

the islands, nor did the ensuing even more opulent millionaire playground period, when the planters, defeated by manumission, had sold out to men like George Parsons of Kennebunk, Maine, who could afford to buy an island (Wassaw) as a present to his wife, as could Thomas Carnegie, brother of Andrew, surprising his Lucy on her birthday with Cumberland. The islands' wilderness was able to contain the vacation accouterments of the very rich of that time. It attracted an exclusive group of New Yorkers—J. P. Morgan, Vanderbilt, Rockefeller and others—who bought Jekyll Island, built The Millionaire's Club thereon, complete with yacht basin, polo field and golf course, and enjoyed the salubrious winters until the biting flies dislodged them at the end of March. To the north some decades later, Ossabaw Island was inherited by Eleanor Torrey West—today the last of the single owners, dedicated to her island's wild state, struggling to keep it intact and in 1977 about to relinquish that responsibility—while the man who was to become the first to end the islands' wilderness, Howard Coffin, Detroit auto magnate, bought Sapelo Island nearby.

A major change of island character prepared the way for his penetration, reversing in a few years the insularity that the long melting of the towering ice cap and centuries of tidal action had brought about.

In the 1920s, coastal Glynn County's population was even poorer than it is today, its tax rolls meager, reflecting the relative backwardness of the South's industrialization. The islands' beauty, fine weather, long empty beaches and open lands might lure affluent new residents, and St. Simons, divided among many owners, could be bought piece by piece, particularly if it became accessible, in those early days of motoring, by automobile. For a brighter, richer future, Glynn County built a causeway across the marsh in 1924, joining island and mainland, an act as reasonable and understandable as Oglethorpe's division two centuries before.

The venture attracted auto-minded Howard Coffin, living the good life on his one-man island, Sapelo. "He was one of that early school of automobile pioneers," a friend recalls, "a lusty group of self-reliant men who didn't know the meaning of 'It can't be done.'" Coffin bought much of St. Simons and all of a long island on its outer edge that he named Sea Island, holdings that now total some 11,000 acres. His lusty self-reliance and the circumstances of his era changed the future of the neighborhood.

Coffin extended the causeway to Sea Island, which he set out to develop for the well-to-do people who owned the first cars. He undertook a total alteration of the slim wild island, replacing its luxuriant wilderness with luxurious shorefront homes, palm-lined streets, neat lawns. Much of the adjacent marsh was filled in and on it rose the super-luxury Cloisters Hotel and the resort's first golf course. "Until a short time ago you were a hero if you filled a marsh," says Alfred Jones, Coffin's cousin, who succeeded him as president of the family-owned Sea Island Company.

Sea Island was to be "a place where we went to live," Jones says, "low density, no crowds. We don't sell, we just let people buy." Coffin went bankrupt to achieve this goal, and Jones, recently retired in favor of his son, became what he calls the "architect of controlled development." Self-imposed zoning, for example, required a minimum of three Sea Island lots for each house, while St. Simons next door was urged to grow into a medium-priced resort, a bedroom for the Cloisters' staff, for people who work in but would rather not live in Brunswick, a town just across the causeway now becoming industrialized, crowded.

The Coffin-Jones vision succeeded for decades, its drawbacks invisible. It was good for the company, good for the country, fitting in with the trend of developing the shore according to private market mechanisms with local political blessings—and no long-range plan. St. Simons' population jumped from a few hundred, pre-causeway, to many thousands and is growing fast, reflecting the population growth of the area, 266 percent higher for Georgia–eastern Florida than for the nation. Almost 2 million cars crossed the causeway in a recent year and the island's 23,000 acres burst with beach hotels and motels for the tourist trade, one of Georgia's top three industries, with second homes, with housing for the retired and for commuters. It is now "going multi-family," a developer says, the phenomenon that already edges most of Florida's coast.

Next door, Jekyll's character too was totally altered. When The Millionaire's Club closed during World War II, Georgia bought Jekyll "for Georgians of average means," built a causeway and bridge to the island, giving us an important sample of state stewardship. Most cause-wayed islands develop fast but on Jekyll the process was even faster than expected because the island is ruled by an authority, appointed by the governor, enabled by law to be completely independent, answerable to nobody, with the power to "sub-divide, improve, lease or sell" a good part of the island.

Jekyll's beachfront is "entirely used up," authority chairman Ben T. Wiggins says, the island crammed with the usual coastal-development array plus a convention hall "to attract groups who would support the hotels and motels." Authority members are officers in the state government, serve without pay, Wiggins says, use revenue from leases for maintenance. Others say Jekyll is a well-known pork barrel, overdeveloped to bring in added spoils. From Jones's view across the sound, Jekyll performs a good function: "With 1,000 motel rooms, a convention hall and beaches, it automatically upgrades St. Simons and leaves Sea Island as it is."

Causewayed islands quicken with the mainland pulse. Cargo to Brunswick, second-largest Georgia port (after Savannah), increased 66 percent in the sixties as the state Port Authority expanded its investment. Seven new industries moved into Glynn County, residents moved out of the city core to the islands if they could. The county wants to continue industrial expansion: "Bigger is considered to be better," one of its commissioners says. "Local office holders are still advocates of 'boosterism.'" The local view is supported by an eight-county Coastal Planning and Development Commission, 75 percent supported by federal funds, that presses for a refinery and more industry on Colonel's Island in Brunswick Harbor. "A policy of 'no growth' is not viable in Georgia" was the stated opinion of Governor Carter's Commission on Planned Growth in January 1974.

On the other hand: "We are very concerned about the future of coastal Georgia," then Governor Carter said in response to a query. "[It] is an important area of the state, both for its valuable natural resources, and the ports of Savannah and Brunswick." Such fence-straddling is everywhere on the coast—laws to protect it, business pressures and people pressures not to. Jones says he was pleased by Brunswick's growth, brought the pulp and paper mill there to raise the standard of living and get the state highway paved, "easier access for Cloisters' clientele." Unforeseen was that this would beget Interstate Highway I-95, which the state hopes will triple coastal traffic in five years and put the islands and environs firmly on the tourist map, an occurrence that now undermines Jones's empire and redesigns still another Golden Isle.

In the 1970s, man's island creation is faced with the

contemporary coastal doomsday duo—erosion and pollution. The beach on St. Simons diminishes; in places there is none left at all; only sea walls, substitute for dunes long vanished under shorefront buildings, try unsuccessfully to hold back the sea. The land erodes; some shorefront homes have been moved inland three times to stay dry. On Jekyll, two hundred acres of dunes, more than half the total, are lost to motels. Floods are frequent as groins and jetties as well as sea walls disturb sand drift. The Jekyll Authority asks Georgia for a "beach enrichment program" to counteract the sandless state caused by the duneless concrete waterfront it authorized.

Erosion so threatened the islands' stock-in-trade that in 1973 Glynn County passed its first coast-protection law, special zoning for beaches and dunes. This rescue effort went into effect on two islands in the neighborhood, but on the third, Jekyll's autonomous authority goes its own way, is considering a plan to "improve" the entire island, dunes and all. Sand, as has been seen, acts by its own laws; repelled on one island shore it is likely to bypass the next as well. Marshes too were lost, built into dry land for the new highways (2,171 marsh acres for I-95 alone) and new industrial sites, used as dumps for dredged spoil or old cars. Eugene Odum discovered marsh value in the fifties and Georgia, moving at its own languid pace, bestirred itself to follow other states and pass a Marshlands Protection Act in 1970.

By this time the great marshes were being attacked on a different front, oil pollution from the sea, industrial and residential pollution from the developed islands, harbors, rivers. The Turtle River, for example, which passes Brunswick on its way to the marsh, has been channelized, like most Georgia rivers, in a program of the Soil Conservation Service and the Army Corps of Engineers to cope with growing inland disposal needs. The river's action is now so degraded that waste dumped into it by such riverside companies as Allied Chemical and Brunswick Pulp and Paper flows raw into the marshes of Glynn, the "vast sweet visage of space" a pre-pollution poet wrote about.

Man's redesign of the coast overlooked the needs of shrimps and blue crabs, 95 percent of Georgia's fishing industry. "All aspects of commercial fishing in Georgia are dependent on the coastal marshlands," the state's Department of Natural Resources says. The crab catch was cut in half between 1963 and '73, the shrimp take reduced in developed counties, oysters—a flourishing industry in the early 1900s—disappearing. For the coastal area cited by the U.S. Department of Interior as "economically depressed . . . worse than Appalachia," the loss is a grave concern. The growing volume of waste in the marshes is everywhere cited as one major cause of the decline; lack of proper management and newly virulent parasites are others.

On a coast that lay peacefully open to conquest, island re-creation seemed logical enough. Step by step— the causeway, Coffin's big buy, Jones's Cloister culture, suburbia on St. Simons, Jekyll's pork barrel—understandable all. Marsh sacrificed to highway, Brunswick industrialized, rivers channelized—necessary all. Or so, with all participants innocent of foresight, it seemed. Now, with disappearing sand, polluted waters, declining numbers of shellfish, the requirements of a functioning coast are somewhat better known, and innocence ends.

But change does not, cannot stop. "We're running out of islands along the Georgia coast," a developer's ad says. The coastal growth that Georgia thought it couldn't get along without is cancerous now, stimulated by taxes, the very reason for building the causeway that started the growth a half-century back.

Time and taxes have destroyed the capacity of all but a handful of the very rich to buy themselves island retreats or even to keep them. In Georgia, as in many coastal states, the economy forces single owners out, hurries development. One criterion only is applied in Georgia real estate assessments, a measure known as *ad valorem,* at its "highest and best use" (economic use, that is, not social or ecological), or what a willing buyer would pay a willing seller. As the supply of seashore dwindles, shore property values soar. Prices are in the $1800 range for one waterfront foot, $150,000 for a quarter acre, a 1976 survey says. Under the *ad valorem* rule, vastly increased taxes result.

Taxes make it impossible for Eleanor West to keep Ossabaw, bought by her millionaire Michigan parents in the early 1900s, left to her, a niece and nephews, whose share she rents. A vivacious, energetic woman, she has tried a half-dozen schemes to maintain intact Ossabaw's wilderness, as well as the red tile roof on the mansion, the boats that wind through the marsh, bringing supplies from the mainland. "I am pushing the panic button," she said in 1976. "I will not be able to support the island much longer without help." Help becomes scarcer as the value of what will be her family's inheritance spirals, even more elevated by the sale of Kiawah, a comparable South Carolina island, priced at $125,000 in the fifties, recently bought for $17 million by the Sheikdom of Kuwait for development by well-known entrepreneur Charles Fraser in a "North African–Arab motif." Considering this the highest and best use of such islands, the county raised Ossabaw's *ad valorem* assessment and in 1975 tripled real estate taxes across the board.

A picture of a smiling Governor Carter with a jovial Mrs. West on an Ossabaw mansion wall, taken on Carter's visit to the island, is misleadingly cheerful. The governor was enthusiastic about the unspoiled scenery, thought it should stay the way it was, but, like most Georgians who elected him, was against state interference with land planning. He would turn the Ossabaw problem back to Chatham County assessors who, he said, should "use common sense" in making assessments of property such as this.

"The county assessor is doing our land planning for us," Jones says and asserts that taxes force him to develop faster than he wants to. He has cut the three-lot-per-house minimum to a quarter acre, says his only alternative is to sell property to some other developer to meet the tax bill. It is ironic that in most instances where the developer is trying to preserve coast resources by building boardwalks over the delicate dunes, keeping densities down, leaving wetlands intact, he courts economic disaster, pricing himself out of the market. The recent financial collapse of Fraser's renowned Hilton Head resort warns any such purchaser to develop every inch and forget the amenities.

A breath-taking example of the results is the International Telephone and Telegraph's Palm Coast, which rips into 92,000 acres of coast down the line in Florida to build a city the size of New Orleans by the year 2000 at the cost of a billion dollars. Encouraged by Richard Nixon during his presidency, ITT now tangles with state officials on the effects that 750,000 more people would have on such essentials as ground water supply and drainage. "They're building a dinosaur, an anachronism," an official says. Regarding permission to build a series of canals, "the tendency is to deny . . ." the state spokesman says, "because such systems age rapidly and badly in urban environments, becoming long narrow septic tanks, discharging low quality effluents into the waters." A nearby resident told a reporter, "The local people don't want Palm Coast, but they'll never stop ITT. There's just too much money there."

Prevented by taxes from island owning, several proprietors would preserve their wilderness domains by transferring them to institutional shelters where they would be put to worthy—and tax-exempt—uses. Certain privileges, perhaps lifetime use, a cooperative game warden, and a bit of immortality, were more appealing to some in exalted tax brackets than the dollar profit. On Wassaw, valued at $6 million in 1969, heirs of George Parsons put aside a commercial sale for a $1 million transfer to the U.S. Fish and Wildlife Service, keeping a "home parcel" of one thousand acres, one-tenth of the island. St. Catherines belongs to the Noble Foundation, creation of its owner, founder of Life Savers, and is used for research.

Such preservation effort is under attack. Cost of the coast, character of the coast, and the ever-stronger urge to be on it, are the principal assailants.

Costs shrink institutional possibilities. Foundations, coming on hard times, can't find the necessary millions. Tobacco heir Richard Reynolds' foundation made part of his island, Sapelo, a state Fish and Game Refuge, and sold the rest to the United States government for an Estuarine Sanctuary, slated to become a field laboratory against which our activities in estuaries elsewhere hopefully can be assessed. For years, Eleanor West could not locate a private or public institution that would spend a reported $8 to $10 million and keep Ossabaw wild. U.S. Fish and Wildlife, for example, which bought Wassaw, owns three smaller Georgia islands, has some 330 such refuges, more than 32 million acres, is constantly pressed to increase its holdings because of the galloping number of endangered species. Then U.S. Assistant Secretary of Interior Nathaniel Reed tells an interviewer that costs make further acquisitions in Georgia or elsewhere "dubious in the extreme." The Nature Conservancy, largest private land-saving organization, spent seven years buying most of Virginia's barrier islands with money from a private trust, now struggles to protect what it owns, and says, "no one else is available with the resources, experience and intent to protect."

One of its more dramatic last-minute rescues was to snatch Ossabaw from the developer's maw in a three-way everyone-benefits coup, a Conservancy specialty, announced in mid-1977 to be completed in May of 1978. 25,000-acre Ossabaw, assessed at more than $15 million, will belong to Georgia for a mere $4 million of state funds. "A proud day for the State of Georgia," Governor George Busbee exulted, promising to preserve "this magnificent portion of Georgia's natural legacy." Mrs. West and her family agreed to a bargain $8 million sale price which would bring them a comfortable tax deduction. The purchase money came from the state plus another $4 million donated to the tax-deductible Conservancy by 87-year-old Atlanta millionaire Robert Woodruff, former Coca-Cola chief. "All parties agree that there will be no bridge built to Ossabaw," a Conservancy spokesman says. But crossing the jammed state-built causeway to overcrowded state owned Jekyll Island gives us pause to wonder if Georgia can resist a repeat for very long.

Such preservation of bits and pieces, a popular crusade, has its roots in the first days of land saving. It is a respectable philanthropy, an applauded public function, but it is *land* saving, a terrestrial idea that may not be transferable to the ever-moving interlaced world where land and water join. Because of the coastal character, no Golden Isle is isolated from Brunswick's wastes pouring into offshore waters, or from St. Simons and Jekyll deflecting sand and polluting the marsh, from the effects of development such as a Kiawah to the north, a Palm Coast to the south, all part of the island matrix which, by its nature, is indivisible.

There is difficulty in preserving the natural system on a single island. "We have no way to examine the effects of pollution under controlled circumstances," a Georgia coast authority says, "because we have no place that we can control."

People's wish for shore and ocean, a great populist push of the seventies, unprecedented, unavoidable, unstoppable, shoves preservation aside. Newly crowded polluted environment, newly available leisure to escape it, and newly dense development of the coast give people more reason, more time and less chance to get to the coast than ever before. Ninety-two percent of it is in private ownership; in Georgia only 5 miles of beach are publicly owned, 23 miles officially available for public use. Sixteen million people are said to live within a half day's drive of the Golden Isles; 20,000 to 40,000 cars a day zip along the new highway, and at least half the occupants are tourists, an estimate says, perhaps 5 million a year, looking for a place to rest, cool off, swim, walk the beaches.

This need changed 40,000-acre Cumberland from a Carnegie fiefdom to a National Seashore, the ninth in the roster of the National Park Service (NPS). Forty years ago there were no Seashores. NPS, just starting to deflect its attention from the west, turned up dozens of east coast possibilities, including the Golden Isles, and bought one, Cape Hatteras. Twenty years ago it entitled an east coast survey *Our Vanishing Shoreline,* and recommended acquisition of at least 15 percent of the coast, surprisingly still available, bought part of Fire Island and a strip of Cape Cod, left Cumberland on its list of desirables. It was not until a Carnegie heir sold his share of the island to the ubiquitous Fraser that NPS, helped to action by $5 million from the Mellon Foundation, finally got the purchase of the threatened island under way.

"Maybe ten thousand visitors a day," a park ranger says, bouncing his jeep through a tangle of palmetto and scrub pine, scaring up a wild brown mare and her colt where one of three projected access roads to the eighteen-mile beach might be. "Maybe two thousand a day," a local developer says, surveying his crucial mainland acres close to the planned Cumberland turnoff from I-95. An exponent of the interchange culture, where the lifeblood is the credit card, he intends to build two thousand motel rooms, an amusement park, swimming pools, "plenty to keep them busy," he says. "With all our facilities, a lot of tourists will never make it to Cumberland at all."

There was a three-pronged motive behind the multimillion-dollar purchase of Cumberland that Congress authorized in 1972 ($10.5 million for acquisition, $19 million to develop park services). It was intended to preserve, in Georgia's Senator Talmadge's word, "the last outstanding seashore area on the Atlantic and Gulf coasts from reckless development" (a goal realists would question, as below), to meet the vociferous demand for "ocean-connected recreation," and to make a positive impact on the economy in the neighborhood.

NPS was to have its troubles achieving this something-for-everybody aim.

There is talk of banning causeways, cars, concessions, of limiting access, of "determined sensitivity toward preserving the fragile natural values . . . while developing visitor use." While planners study, state authorities exert pressure for facilities that will bring the county more taxes, a thousand new jobs. Postponements follow delays; five years after the long-planned purchase, no request to Congress has yet been made for money to begin.

Part of the trouble is the use-*vs.*-preservation schizophrenia that torments NPS everywhere, worse, now, than ever. In the last twenty years the number of parks has increased from 181 to 286, visitors from 50 million to 227 million. The staff-to-visitor ratio has been reduced from 1 to 27,000 in 1960 to 1 to 44,000 in 1975. "It all comes down to manpower and money," NPS Director Gary Everhardt, a former ranger, tells a visitor. Secretary Reed blames Congress: "It was more fun to authorize the parks than the people to take care of them." Manpower and money lead the ranks of a multitude of NPS difficulties rounded up in a spate of recent investigations, such as the *Wall Street Journal*'s "Wide Full Spaces."

With the acquisition of Cumberland, NPS stumbled on an added and profound trouble, a teeming hornet's nest at the heart of the new tough-minded competition for the coast. The mild-mannered ranger-run bureaucracy is patently not equipped to battle the swarming, profit-minded pressures that in fifty years transformed the Golden Isles and much of the east coast and grow more insistent than ever. At older Seashores, conflicts now erupt. Two bridges to Assateague bring a million visitors in 1975 to that Seashore opened twelve years earlier, a thirty-seven-mile barrier island off the Maryland-Virginia border, threatening the famed Chincoteague oysters and wild island ponies with their campers, over-sand vehicles, their very numbers. Meanwhile, groins on the jammed Ocean City shore to the north trap and hold sand that would naturally nurture Assateague. It erodes. On summer weekends, 30,000 people crowd onto New York's Fire Island, the thin 31-mile sand reef where building has increased 40 percent since part of it became a National Seashore in 1964. "Crowded cottages and cesspools on the narrow strand, boatels on the fragile marshland" are noted in the press; angered residents charge NPS in the U.S. District Court with "irreparable injury" to the island, file suit against the Department of Interior for failure to enforce the law.

There is no sign that the newest Seashore—stillserene Cumberland—can escape. On the contrary, its tripartite intentions, by definition, lead it straight to the fate of its slim sisters along the Atlantic seaboard.

There will be outbursts, protests, efforts to stop the clock or at least slow it. Islands still arouse fervent feeling, danger to islands stirs initiative. Down east in Maine, where islands are smaller than in Georgia, the benevolent, one-man-one-island ratio has survived somewhat longer. Mrs. David Rockefeller believes it should continue. A two-island owner who likes to sail with her husband in the bay past the exceptional beauty of the wild islands, she saw seven houses where there had been one, three new docks on a formerly pristine, rock-strewn shore. A brisk, determined lady, she set out to establish a system of voluntary conservation easements. These consist of giving away land-use options—no further buildings, only one more dock, deck, house, or whatever limitations the owner chooses—to appropriate public or private agencies to become a permanent part of the land's title.

"The time hasn't come yet to stop private ownership," Peggy Rockefeller says. "People have a right to own land and protect it." The protection she chose is one-dimensional, terrestrial. With the help of the family lawyer, who worked out a similar scheme for Pocantico Hills, and island-owning friends ("They come here for island solitude, wilderness"), she saw an enabling act through the Maine legislature in 1970, founded the Maine Coast Heritage Trust in 1971. In four years it collected 108 easements "protecting" 12,000 acres, 57 whole islands, 6 parts of islands. The protection is from development, leaving what happens to the rest of the island matrix to take its course. "The Trust has survived the Rockefeller Handicap—preservation for the wealthy," John Coles, outspoken editor of the *Maine Times* says. "It appeals to traditional Maine independence."

The same Yankee independence reacted in rage to Senator Edward Kennedy's public-sector effort to keep Martha's Vineyard, an island off Cape Cod, from total development. In the early seventies, the hundred-square-mile island hovered on the edge of transformation from a rural fishing and farming community to a tourist–summer-people economy that was about to slice up the land, destroying rare geological, historic and natural phenomena. Kennedy proposed to Congress a unique Island Trust that would put the Vineyard and neighboring islands under a federal-state-local protectorate, a measure fiercely opposed by the local island power clique. The modern Yankees' interest was economic, parochial, firmly dedicated to short-term profits, and powerful enough almost to kill the Kennedy bill in committee in 1976, while honeycombed housing and standard tourist attractions transformed the Vineyard into Anyplace, U.S.A.

We begin to expect that such shoreline change is inevitable. At the same time its end product terrifies. You can see it towering into the sky around New York Harbor, the largest urbanized piece of coast in the world. Within a 2-hour drive of it, 10 percent of the nation— one in every 200 people on the planet—crowd together, and islands are so transformed, concretized, built upon, that they seem firmly part of the main. Children and adults in the core cities of this New York/New Jersey coast, unable to escape, are turned more savage than in any previous civilization on the shore once admired by Verrazzano for its wild flowers. Many are poor, as underprivileged as Georgians in whose interest coastal growth is currently undertaken; their murder-rob-rape syndrome fills island-city streets with fear. They have no chance to know or cultivate the gentling rhythms of the coast they live on.

For these people, these cities, this fear, the United States attempts to *undo* growth around New York Harbor, helped along in this innovation by the existence of surplus shore-front military forts ("They're not expecting Spanish galleons any longer," a planner says), abandoned airports, publicly owned beaches, an incipient housing project. These have been collected into the Gateway National Recreation Area, a package of four separate parks around the harbor, some 26,000 acres with more than 9 miles of ocean beach, 12 miles of bay beach, wetlands, marshes, even a wildlife sanctuary built on a dump. *Spartina* grass struggles up in a marshy place among disintegrating mattresses, old tires, junked cars; empty buildings, vandalized to their skeletons, line a plastic-strewn beach. Eighty percent of all ocean dumping in the United States goes into the waters of the New York Bight, including one and a third billion gallons of sewage each day (one-third untreated). "It is," marine scientists say, "a highly stressed environment."

Neglected and stressed as it is, this almost empty coast exists within hailing distance of Times Square. It has become the first national urban park (a second, Golden Gateway, is under way in San Francisco) by massive concurrence, the support of a campaigning President Nixon ("We must not allow our cities to become concrete prisons"), of several successive Secretaries of Interior, of state and local politicos in 22 counties and 550 municipalities, of public and private planners, of citizens. It has taken decades for it to be assembled, approved by Congress, open in part to 5 million visitors in 1974, 7½ million in 1976, less than half the potential although it is more by far than in any other NPS unit, as is its $6.9 million annual budget. It will cost the nation a whacking $300 million to carry out its intentions for high intensive use—65 square feet a person, 20 million people a year.

Despite money and enthusiasm, transportation to Gateway is hardly available, highways and subways are jammed to capacity, a ferry scheme discarded as too costly. No way has been found to get the people for whom Gateway is intended—the 34% who do not own cars—to its shores. Or to make it safe for them, once there, to follow their primal instinct and get into the water.

The city in all its vastness comes first. There has been no change in priorities, and without it, human alterations are not easily undone, even for the universally applauded objective of bringing an island, a coast, back to life.

7
THE PLUNGE

The thin edge was cut into a new design by the massive
people pressure of the past half century. Now an ever-
expanding megalopolis uncoils along the Atlantic, multiplying upon itself with unprecedented haste. It brings us
to an even more daring development. Having crowded
the coast to capacity, we are poised to pierce a frontier
beyond our fathers' imagining. We would grow out into
the water.

Penetrating prehistoric rock on the sea floor might
find us oil, building on the sea surface could give us
space for the neighbors nobody wants, and by some
magic the sea may, after all, prove the desperately
needed answer to the immediate dilemma of waste. We
fantasize further spectaculars—floating airports, parks,
even cities, undersea hotels that would send taped music
out to resound against the music of the deep—and, as we
visualize, we put facile minds to work to make such science fiction come true. Carried away with images of how

the ocean might be transformed, some call the approaching era the Blue Revolution.

There is a superb technology that can effect this plunge but it is yoked to a puny, immature mate—our undeveloped perception of the coast that must bear the weight of this change. Know-how races ahead of knowing what effects will be, eagerness for a whole new world for growth leaves far behind the frail beginnings of understanding that world.

This incompatible coupling caused us to build on dunes—and flood lowlands as a result, fill in marshes—and diminish marine life, drive sand off beaches, overkill shorebirds, overfish. With the move offshore, we continue to overlook what we know and what we don't know but could pursue. Impatient, the Blue Revolution sets aside man's capacity to learn the requirements of a functioning coast and instead encourages the short-sighted if brilliant ability to conquer new territory. Its siren song is seductive—expand, expand—its sponsorship rich and influential, its promise exciting, making it easy, as one scientist puts it, to ignore the importance of ignorance.

This philosophy of conquest is encouraged by the conviction that it will boost the nation's economy. If there is oil in the east coast's Outer Continental Shelf (OCS) as suspected, the black gold can bring prosperity. If offshore waters are used to cool the awesome heat engendered by nuclear power plants, there will be more energy, and even more prosperity. Seaward expansion in general has an inviting balance sheet. The economic potential of U.S. offshore resources will reach $34 billion to $45 billion by the year 2000, according to Senator Ernest Hollings, acknowledged congressional leader in ocean-coast matters. The attraction of an improved fiscal future and its quick, dramatic achievement entices us to the move offshore.

The progress in a few short years astounds. In the 1960s and early '70s we were engrossed in converting the coast to accommodate a tourist economy, housing developments, industrialization. In 1976, drills whirled down into OCS rock off New Jersey and in Georges Bank, the fabled fishing grounds east of Cape Cod. Test borings are financed by a thirty-one-member consortium, ready to dig wherever oil is discovered; actual production could begin in 1981. Already, supply ships and drilling rigs assemble in a Rhode Island shipyard; "Atlantic oil is going to be an enormous operation and we plan to be at the center of it," a state official says. So too does Tar-Off, an impregnated paper towel for cleaning blackened soles; it's company started manufacture in 1975, finds sales ahead of expectations in Florida and California, will go where oil goes.

In Jacksonville, Florida, Offshore Power Systems, a Westinghouse subsidiary, prepares to build eight floating nuclear power plants. Its president was pleased that President Carter stressed the need to standardize design and cut production time, says "When he talks about doing that he's talking about us," and despite vigorous citizen protest, plans to lay the first keel late in 1978, using a $13.5 million crane, the world's second largest. The first two plants, ordered by New Jersey's power company, will make up the Atlantic Generating Station, to be completed and floated into place a few miles east of Atlantic City by 1984, OPS says, and to start producing power as soon as it is connected to the underwater cables on the sea bottom. In 1975, offshore islands were the subject of a two-volume evaluation by the National Science Foundation, which identifies the most likely place (middle Atlantic region) and use (oil refineries, because they get the "most rejections" for on-shore sites) for the nation's first artificial islands. In the same year, Aquapolis, 19,000-ton prototype of the floating city of the future, was the Japanese government's display at Expo '75.

Chances of our surviving will be fewer if the Atlantic pounds against floating nuclear plants off New Jersey, or anywhere else. At best these plants will run the same risk of accident as those on shore (many experts say the potential is greatly increased by the proposed offshore siting) but with a crucial difference. A core meltdown into the land has the possible protection of the much-discussed "glazed-earth reaction" that might insulate the deadly radiation, keep the disaster local. The same accident in water would have no such chance. When molten white-hot radioactive materials hit cold sea water, radiation would contaminate thousands of cubic miles of ocean ("an environmental nightmare" one expert comments), would be released into the air in steam, come ashore with bottom water, move up and down the coast with littoral currents, enter the marine food chain, and such materials as plutonium, which irradiates bone and alters genes, would continue in the environment for centuries.

The raging nuclear debate is beyond this consideration of the coast. In our context here, power plants serve to confront us, point blank, with the danger, one of the most extreme we know about, that can come from moving offshore what is onshore without giving coastal character the number-one priority in decision making. If an accident happens at sea, that character will claim priority for itself, causing New Jersey to share with the world this realization of its death wish.

"The marine revolution cannot be halted," Elizabeth Borghese says in *The Drama of the Oceans,* proposing that population pressure, food shortages, near exhaustion of land-based resources "push man into the sea, his last frontier." Many like-minded enthusiasts would have us believe that there is no choice, that we can safely ignore what we do not know but need to know, reject what we do know that might hold us back from this daring conquest. Decisions, and the millions spent to implement them, appear part of an overall scheme.

But there is no scheme, only a momentum, accelerating toward the plunge.

Dangers of this plunge—those now in sight—are severe enough to rouse curiosity about the decisions that have brought us to the brink, decisions that are extraordinary in light of unfolding knowledge that, even in its embryonic stage, lets us anticipate what we are now tempted to buy blind. Dissecting some of these decisions reveals that the nation, in its nonstop push for expansion, disregards this ability to anticipate, to realize the staggering price of what it is doing.

To be able to look ahead on the coast is a new achievement. In 1890, for example, and for decades thereafter, the ocean was thought to be the ultimate sink, the supreme dilution basin. It seemed convenient, inexpensive and agreeably out of sight to dump wastes from the New York–New Jersey area into the New York Bight. Troubles that would stem from this decision were impossible to foresee then. No one imagined that in eighty years wastes would fill up the submerged Hudson Shelf Valley, form substantial hills on the ocean floor and a great black mass of sludge, so virulent and vast that natural ocean processes can no longer decompose it as fast as it is dumped.

But basically, in the 1970s, despite the new information, we make the same decisions our forebears made without it, and the longer we cling to the ocean-sink concept, the worse the impact is. Thousands of pounds of PCB, a highly toxic and carcinogenic member of the chlorinated hydrocarbon family, were dumped into the wide and vigorous Hudson River by two General Electric plants. Absorbed by marine life from plankton to migrating striped bass, PCB concentrates as it moves up the food chain from smaller to larger creatures, its effects "at worst, fatal, at best, unknown," a study says. The river was closed to fishing in 1975 and the dumping was stopped, but that could not prevent aquatic life from passing the compound along in Atlantic waters. PCB is now "a pervasive pollutant of air, earth and water," swallowed by infants in their mothers' milk in ten states.

Similarly, it was something less than a century after dumping had started that the Black Mayonnaise threat to local beaches made headlines. In 1974 a marine biologist said it was a finger of sludge creeping shoreward; a regional official of the Environmental Protection Agency (EPA) said that was "a lot of hogwash." EPA considered moving the sludge dump further out to sea or up the coast, but protests and increased costs of barging were said to have killed the idea; the mayonnaise went away and the dumping continued at a faster rate than ever.

In 1976, waves of stinking debris—raw sewage, tar, grease—washed up on almost one hundred miles of Long Island's south shore during a warm week in June. The beaches were closed, New York Governor Carey declared the region a disaster area, making it eligible for state funds to clean up the mess, President Ford sent one hundred Job Corps trainees to help. Angered townships sued each other, the county, New York City, but there was no consensus on where the sludgelike tide came from, or why. It was two weeks later that the fishermen found the huge fish kill on the ocean floor just south of the sludge dump site, the brown water that befouled the ocean, the oxygen below the minimum required for survival.

It happened too soon to be completely understood. There is still only scattered, if suggestive, information. In a sample of mackerel eggs collected in the Bight, a surprising third had abnormalities; among 9300 visitors to Coney Island beach one summer weekend, the incidence of gastrointestinal symptoms (vomiting, diarrhea, nausea) among swimmers was significantly higher than among nonswimmers. But research has made no real strides; without facts there is no way to prevent whatever new disasters may appear because the ocean dump, having never been at this stage before, is writing its own script.

Nevertheless, we make only ad hoc adjustments when trouble shows up—stop dumping PCBs in the Hudson, look for another sludge site for the New York area—and make curiously naïve efforts to identify and punish a villain, in the hope, perhaps, that some one town, company or person is responsible for the trouble. Thus, Long Island towns sued each other, and in the same Bicentennial year, the largest fine ever imposed on a corporation or individual for polluting the nation's waters—$13.3 million—was charged against the Allied Chemical Company by a Federal District Court for knowingly dumping deadly Kepone, a nonbiodegradable chemical cousin of DDT, into the James River and Chesapeake Bay and thence to offshore waters.

From the Gay Nineties perspective, using the ocean as a sink was as logical as shooting shorebirds, and stood unchallenged until the sea itself rose up, Black Mayonnaise in hand, and made its threshold known. "Contamination of the ocean has begun," Edward Wenk, advisor on marine affairs to three Presidents, says in *The Politics of the Ocean*. Symptoms such as the abnormal mackerel eggs and vomiting swimmers begin to mount; empirical scientific evidence accumulating everywhere suggests that there are limits to what the ocean can take and keep functioning, and that we are approaching these limits at a fast clip. Cancerlike growths substantially increased in oysters, clams and mussels from polluted bays and rivers, six independent 1976 studies discovered, a finding regarded as a primary indicator of contaminated conditions. In May, 1977, federally unacceptable pollution moved further eastward along Long Island's shore, closing 2600 additional acres of clam beds; more closings were expected. In July, on the hottest day of the year, 50,000 bathers had to leave the water when a huge mass of sewage debris floated ashore along two miles of South Shore beaches. Descendants of the nineteenth-century dumpers-in-ignorance can anticipate more serious trouble and some, half-heartedly, have set about quantifying the effects of these undersea hills and sewers, looking for a technological fix and for alternatives, but without great optimism. "We do not live in a world of zero risk," the EPA man says in remarkable understatement. "Handling sludge must have *some* environmental impact."

Sometimes solutions exist but get shoved aside in the hustle. The striped bass that escape PCB poisoning in the Hudson, for example, are exposed to the heat of six power plants on that river that will cut the bass population to half its pre-1974 levels. A closed-circuit cooling system would stop this heat pollution; "Con Edison . . . has fought closed circuit cooling systems each step of the way" and power companies up and down the east coast follow suit, the Natural Resources Defense Council reports. Sometimes solutions might be invented if the necessary facts were assembled, as in the matter of sludge, or at least the facts would indicate whether a solution within the present framework is even possible. But studies, fact collecting and analysis are slow to get under way, limited by lack of funds, facilities and priorities. The U.S. Office of Technology Assessment, reporting to Congress on the multiple demands of studying what's happening to the ocean, schedules its Marine Disposal of Wastes study to start in 1977.

The recently acquired ability to anticipate what's next lies fallow. Invention of new ways to grow diverts attention from the unpleasant results of growth—more waste than we know what to do with. We avoid the question of where to put our mountainous discharges, our billions of gallons of throwaway poisons. The earth's air envelope is close to its tolerance level; there's not much room left on land. There's not much room in the ocean either, but apparently we have to wait for proof absolute, however horrendous, to convince us that it is past time for an all-out attack on the problem, a major thrust to find a new solution. Without it, we keep dumping, keep gambling.

The stakes, high as they are, are even higher when there is oil at the end of the offshore rainbow. There's more to gain, more to lose. To extract, transport and refine east coast oil from the Outer Continental Shelf will change the U.S. energy picture for a few decades at most. With the present "state of the art," it will change the future of the still viable, still surging east coast for centuries.

Money investment alone suggests that without a fairly fundamental cultural change there will be no going back. Multiples of millions have been spent by industry to prepare the engineering, documentation for government consent, preparations to dig or to build. One of many exploratory oil digs, for example, cost $31 million, one year of planning a floating nuclear plant $23 million, one plant itself some $600 million. The nation spends more millions investigating, licensing, establishing its OCS rights and, in 1976, put a billion dollars into a "negative impact fund" for east coast states, to ameliorate whatever difficulties offshore oil development brings, a not inconsiderable carrot.

Beyond money, the investment is in a state of mind —frustration with the confining troubles of the crowded coast, an eagerness for more space, more energy, more jobs, more profit, for getting on with it.

Against such all-American pressures is the counterweight of coastal character. The scenario of the offshore move does not take into account the flowing system's essence, its indivisibility. The intricate connection between water, sand, fish, marshes and the rest—a coastal absolute—when first discovered appeared to be a natural marvel, a phenomenal design incorporating the most delicate balance between a dozen different systems into one that produced, protected, and encouraged life.

Now this same unity terrifies. The system's essence becomes its threat. Newly added pollutants will not be isolated but show us, insistently, over and again, that what happens to one part of the coast affects the rest. Even before the plunge offshore, the coast demonstrated its indivisibility, almost as if it were issuing its final warnings, giving us a fearsome lesson, another chance to understand how it works before we undertake the bold, uneducated move into the water.

Oil, sometimes lethal, sometimes less, is never a healthy diet. Existing knowledge about oil in the marine environment is "at a very primitive stage," a Ford Foundation report, like dozens of recent surveys, says. The immediate catastrophic local effects of oil spills are known to a degree, helped by the fact that two huge spills took place almost at the doorsteps of oceanographic institutions, one in England, one in Massachusetts. Still mysterious is the new condition brought about by unprecedented amounts of oil in the water, and by the workings of the stubbornly indivisible coast.

Most serious analysts say our rudimentary understanding of the effects of oil is clouded by the emotionalism of conservationists, determination of industrialists, partisanship of politicians and rivalry among them. All agree that more information is essential. But there are now some observable facts.

Shore and sea birds are attracted to oil-slick water, some say because from the air it looks like a mass of delectable silvery minnows. When their feathers become coated with oil, birds lose their insulation and freeze to death in any season. If they dive directly into the oil and swallow it, they choke to death. Most of a migrating species, collected in a flock, can be simultaneously exposed to the dangers of an oil spill; if it happens at breeding season, it can eliminate the population. The 31 million gallons of oil from the immense *Torrey Canyon* spill killed 40,000 to 100,000 birds—20,000 of them guillemots, a species of seabirds, and 5,000 razorbill auks. In the Delaware River, oil pouring from two colliding tankers engulfed the crowd of small diving ducks called ruddies that were wintering there. "Saturated with oil, weighted down by its density, they are drowning, starving, and being pecked to death by emboldened sea gulls," an observer reports. Of the 12,000 in the river at the time of the collision, 8,000 died.

"The most effective control procedure to limit the contamination of birds by oil," according to the American Petroleum Institute's 1974 research, "is to patrol the slick and to direct rockets and alarm sounds at the spill area." A Fourth of July celebration at their accustomed haven would be a considerable change for migrating birds, ready to rest after thousands of miles en route. No one has ventured to guess how long it would take to achieve the genetic adjustment that would produce oil-

proof feathers and gullets, rocket-proof nervous systems, or whether such adjustment could evolve before
decline and extinction, the one reliably fast reaction we
know something about.

Oysters, clams, mussels—filter feeders—are easy oil
victims. The initial kill from the Northern Gulf spill in
Maine was 33,000 pounds of lobsters, almost 3 million
pounds of clams, followed by at least 10 years of hydrocarbons in the sediments (and clams) of the area. Concentrations as low as 10 parts per million can kill
shellfish, one part per million reduces oysters' growth
rate.

One part per million—it would be an ounce of ver-
mouth to almost 32,000 quarts of gin—is a deadly drink
for fish eggs and larvae. In mature fish it has a less than
lethal but nonetheless catastrophic effect. Certain oil
components that dissolve in water inhibit the fishes'
chemoreception centers, destroying chemical communication. Penetrating the cell membranes of anadromous
fish like striped bass and salmon, oil interferes with the
homing mechanism, destroying the poignant "place
faithfulness," and causes random nerve responses, one
scientist says, "which set fish into erratic swimming activity that may culminate in death."

But oil has entered the biosphere to stay, scientists
say. All parts of the coastal system are now submerged
in a never-ending oil bath that penetrates the cells of fish
and shellfish, smothers *Spartina* grass in the marshes,
kills fish eggs, larvae, barnacles; each death in turn
affects the watery world of which it is a part. The petroleum hydrocarbons are notably persistent, stay around
for years, decades, perhaps longer, time for natural dis-
persion to produce "oil droplets of particle size," Dr.
Max Blumer of Woods Hole Oceanographic Institute
observes, easily ingested by many marine organisms to

accumulate eventually in food as carcinogenic compounds, long-term poisons.

The more oil we spill (and oil wells will contribute far more than tanker spills and other present sources), the more forcefully does the integrated coast character assert itself, and the more offshore conquest takes on aspects of Russian roulette.

Oil will suffuse the coast near which wells operate, according to the computerized calculations of the U.S. Geological Survey. A preview made headlines in the winter of 1976 when the tanker *Argo Merchant* broke up on Nantucket shoals, spilling some 7 million gallons of heavy crude oil almost directly over the Georges Bank fishing grounds, the tenth largest spill ever and perhaps the worst of all in effect because of its location. The next day, hundreds of oil-soaked birds washed ashore on Nantucket's beaches, deep-sea scallops twenty miles away had oil in their shells, there was fear for floating eggs and larvae of spawning cod, haddock, flounder, for bottom feeders and lobsters in the deep canyons where oil might settle. Fishermen sued the tanker owners; Governor Michael Dukakis, a staunch supporter of the offshore oil dig, asked the President to declare the southeastern part of his state a disaster area.

This same Georges Bank area, along with the Baltimore Canyon, has been chosen as the immediate oil-producing site. An oilfield there will have a 20-year life, during which there is a 91 percent chance of at least one huge spill (over 1000 barrels), 99 percent chance of at least nine 50–100-barrel spills, and more than 1700 so-called nickel-and-dime spills. The big spills have a two-thirds chance of coming ashore; even without them, oil from the field will cover the beaches on nearby Nantucket Island with from 3.5 ounces to 2.2 pounds per yard, will deposit somewhat less on Martha's Vineyard and Long Island, the Survey says. The natural resources it believes especially endangered are offshore spawning areas, the wintering area of ducks, particularly the American eider (another preview—eider population was decreased from 500,000 to 150,000 by a winter tanker spill in the same place 20 years ago) and the beaches that, in addition to their threatened big doses, will suffer "chronic fouling by small-scale spillage." "Don't knock it," a politician says, "it's the smell of jobs."

We are smart enough to invent a way to take oil off feet but not to keep it out of our innards. The ocean will have more oil hydrocarbons in its solution than ever before. No one knows what the effects will be; research doesn't go fast enough to find out. If there is a safe level of oil in the water, no one can say what it is. "There is little data to enable us to determine impacts of oil toxicity," a marine scientist testifies, and the Council on Environmental Quality says that, in the absence of facts, it bases its assessments on probabilities, even informed guesses.

Wrong guesses can have lethal results. An MIT group studying the probable impact of the Georges Bank dig looks at floating oil near soft-sediment areas like salt marshes, says it will penetrate and stay for at least ten years, disastrously affecting flora and fauna. Until very recently, compounds that would sink oil were used for spill cleanup, with the thought that, in addition to the out-of-sight, out-of-mind advantage, ocean action would dissolve the drowning oil, keep sediments and marshes clean and unaffected. Now we find out that it was a poor gues. Then a certain part of petroleum sinks, Woods Hole scientists have discovered, it spreads on the sea floor, where it forms "an asphalt-like surface" that keeps its toxicity, kills all but the most resistant organisms and plants, leaving an underwater desert like "the Great Dust

Bowl," susceptible to erosion and the spread of pollution.

Liberated OCS oil can thus turn marshes and sea floor into a perpetual poison bank—its dividends, hydrocarbons that ooze into the marine atmosphere to wreak their vengeance. Oil accouterments—pipelines, access channels, oil rigs—will do their share in altering offshore waters, interfering with its inhabitants, and those humans who would catch them.

Oil will change the shore where it is brought to be refined and distributed. Consider Louisiana, the supreme oil coast state, where OIL FEEDS MY FAMILY is the popular bumper sticker and 90 percent of all offshore wells in the United States are drilled. This Delta coast no longer gains wetlands as it has done for 4000 years, but has lost some 500 square miles of them. It now experiences a petrochemical boom that increased the coastal population by 51 percent in 20 years, built 100 major petrochemical plants, bringing $5 billion into the state and causing 8,000 miles of oil pipeline to stretch across the disappearing marshes. "Many mistakes have been made in Louisiana, including serious disruption of marsh ecosystems," Lyle St. Amant, state coast official says, "primarily because oil production . . . operated in an unregulated fashion for nearly 20 years."

An oil coast is being born in Stavanger, a thousand-year-old Norwegian fishing village not far from the colossal new North Sea oil find. "The first flush of economic success is obvious everywhere," a visitor reports, "everyone drives fancy new cars, dresses beautifully. Even the taxis are new Mercedes." Beer costs $2 a glass, a smorgasbord spread at the new oil-financed hotel, $70 per person. Although Norway doesn't need to extract oil in a hurry, it is pressured by the oil industry, seduced by visions of billions of kroner and a higher standard of living. The traditional fishing industry declines, suffering interference from oil impediments that litter the ocean, may leave once-quaint postcard Stavanger completely dependent on oil, an adjustment that will painfully change an old culture. The change may be accomplished just in time for the oil to run out.

And run out it must. Oil and its companion, natural gas, once thought to be constantly produced in the earth, are now known to be nonrenewable resources. Animal and vegetable matter in the sea, piled up for hundreds of millions of years in the oxygen-deficient environment of what was once the ocean floor, was compressed by layers of sediment, underwent interminably slow chemical change to eventually form the oil compound. Hidden in thick sedimentary rock, folded into the globe, oil migrated upward until trapped by certain large rock pockets, where it was only recently discovered by man.

A great deal of the planet's oil has been used up since. "Children born in the last ten years will see the world consume most of its oil in their lifetime," King Hubbert, eminent geophysicist, said in 1976. Edward Teller believes the era of fossil fuels will be exhausted in the first years of the twenty-first century, and Senator Hollings comments in respect to these fuels, "We are entering the twilight of our productive years."

During these sunset years the world will divide what's left, possibly 2000 billion barrels to 2500 billion. Of this the United States has 150 billion to 160 billion barrels, the U.S. Geological Survey says, although it could be anywhere from 72 billion still-hidden barrels to 400 billion, depending on who's doing the estimating. Most agree that a third of this will be found in the OCS, from 5 billion to 13 billion barrels of it off the east coast. Recovery of this fraction might slightly reduce the 40 percent proportion of oil imported for U.S. needs (at

1976 rates), might supply the energy-hungry east coast for a while. Then it will be gone.

Offshore oil is a short story for all the money made and damage done. In less than a century we invented the skill to find it and extract it, and in another century, according to predictions, we will exhaust it. The first U.S. offshore well was drilled from a wooden platform in California waters in 1894, the first out-of-sight of land off Louisiana fifty years later. By 1950, somewhat less than 2 percent of the world's petroleum came from offshore wells. Then the technology took off. By the 1970s, there were 17,000 wells off the shores of California, Alaska and the Gulf states, dug as far out as 1500 feet of water, with predictions that it would be 6000 feet before 1980.

The risk we were taking was hardly known until Well A-21 blew in the Santa Barbara Channel, gushing wildly for ten days to pour five million gallons of oil into coastal waters, the whole terrible disaster seen by the citizenry everywhere on television and powerfully recorded in Robert Easton's *Black Tide*, which calls it "the blow-out heard around the world." It aroused infuriated Californians to action in GOO (Get Oil Out) and spurred oil-company efforts toward new safety measures and the invention of possibly safer submersed platforms to be used on the ocean floor.

"Sophisticated . . . the final answer," Elizabeth Borghese says, encouraging this new Blue Revolution technique. Larry Hall, Maine environmentalist, calls it "oil's obscene industrial party . . . a weird sociological suicide."

The importance of getting oil from the east coast OCS depends on the perspective. It is not within the purview here to embrace the pros and cons of the U.S. energy policy but rather to distinguish this segment of it from the long view of coastal function. We have converted the east coast from the wilderness, source of a good life for early colonists, to the barely functioning crowded coast we know. We have constricted our freedoms—for living on the shore, harvesting food free of poison, enjoying sand and water—but can still know such simple marvels in some places. To dig for OCS oil we now begin to convert the coast again, this time to a petrochemical economy, with or without Mercedes taxis, but with—no question about it—unprecedented amounts of oil in the waters, marshes, sands.

It is startling to discover that the question—to dig or not to dig—has never been openly decided. It slipped into place, borne by a dozen lesser decisions.

Ever since offshore oil became a gleam in the industry's eye, a dispute has raged over the potential profits hidden under the waves. OCS exploitation takes place on public property, so the spoils are divided between public and private sectors, and by 1920 there was a move to get a percent of the potential wealth for the United States. The states claimed the seabed off their shores for their own; the nation believed it the property of all the people. Their battle would have wrinkled the noble brow of Hugo Grotius, whose statue looks seaward from an ancient Delft square where more than three centuries ago his great work *Mare Librum* envisaged freedom of the seas for all to share. The fascinating history of man's claims to the ocean ever since swirls around economics; so too does this controvery between the nation and the states.

The issue is fiscal—not whether to dig but who will get the royalties. President Roosevelt and Secretary Ickes tried to take ownership for the nation of the offshore California seabed in the thirties but Congress did not agree, and the matter awaited President Truman's use of executive privilege for his 1945 proclamation that declared the United States ultimate owner. In the fifties

Congress passed a law that would increase federal revenues from offshore oil, giving authority for its development to the Secretary of Interior, who has the power to lease lands, regulate operations.

There was no decision whether to dig or not to dig.

Thunderous challenges attacked U.S. possession; assorted litigations held up President Nixon's plan to step up U.S. oil production, Project Independence. Oilmen, wanting every twilight minute to count in dollars, would hurry OCS digging along. They came out solidly for U.S. oil self-sufficiency, for which, a Tenneco vice-president testified, "it is essential to open the frontier areas of the Outer Continental Shelf." A Chase Manhattan Bank publication pronounces Project Independence goals "vitally important" for the United States, threatens interfering environmentalists with "a backlash leading to an ugly conflict" with consumers, who, it says, would surely win.

Waiting for the U.S. Supreme Court to decide the matter of ownership in U.S. vs. Maine, Nixon had his Council on Environmental Quality (CEQ) make an environmental assessment of OCS oil and gas. Before the report could be written, he tripled to ten million acres the OCS lands that would be made available to oil companies for lease. In 1975, armed with the CEQ report that, not surprisingly, concludes that digging and protecting the coast "are not mutually exclusive" and the Supreme Court's 8–0 affirmation of U.S. rights to the seabed, Nixon had his Secretary of Interior send out the call for leases. By October 1976, the United States had accepted bids from oilmen in a total of $1.13 billion for 101 tracts.

Still no decision whether to dig or not to dig.

Each challenge, each study, each assessment, deepens the nondecision to proceed:

An oil spill risk analysis is made by Interior's U.S. Geological Survey "to determine relative environmental hazards in different regions of the North Atlantic OCS lease areas."

Not if, but where.

A National Ocean Policy Study on Nixon's acceleration program questions implications of "rapid development of OCS oil and gas."

Not if, but when.

East Coast Governors, almost unanimously, attempt to have exploration permits separated from production permits.

Not if, but how.

Each relates to choices *within* the big decision that the public never made. Each choice, each lease, each dollar spent makes the nondecision firmer.

The oil industry, perhaps finding this gray area less secure than it would have liked, embarked on a multimillion-dollar advertising campaign that said offshore oil drilling is safe, clean and nothing to worry about. "The day of offshore oil is a long way off," an Amoco ad said in 1975. "Our job is you." Continental Oil assured conservation-minded consumers that "wildlife and oil production . . . can thrive alongside each other," noting Aransas, where, it said, Conoco and rare whooping cranes winter together. The industry avers that it spends $3.3 million a day on environmental protection and at Christmas, 1976, put up a fund to replace damaged equipment for Massachusetts fishermen who run afoul of Georges Bank exploration. Along the Baltimore Canyon a $2 million seismic vessel, probing for oil, covered its Shell insignia "to keep a low profile until we can explain ourselves," a spokesman said.

Columnist William Shannon needs no further explanation. "Those oil rigs seeking to move up the Atlantic
coast are not the agents of fate or the national interest,"
he says in the *New York Times.* "They are only propelled
by the mindless greed for profits. . . . It is time to call a
halt." To cartoonist Herblock, some months later, it
seems too late. A cigar-smoking oilman and a bureaucrat
laden with offshore leases pose in front of his cartoon's
sign that reads, "CLOSING OUT SALE: THE U.S. ATLANTIC
COAST AREA."

New scenery is on the way to the Atlantic horizon, a
hedge of artificial islands adding their strange shapes
and concrete walls to the towering oil cranes and low-
lying nuclear plants. There are so many applications now
in so many Washington offices that, barring bold action,
some will surely come into being before the century is
out. In the aura of segmented decisions about the shore,
a strong case can be made for industrial islands, for the
Single Point Mooring, a single buoy or tower structure
to which any size tanker, even the monster VLCC, is
moored by its bow. Invented in 1959, 130 SPMs have
been installed throughout the world, 16 off tiny Japan,
none in U.S. waters. Refineries too are needed, an estimated 45 more by 1985, and there's an eloquent case
for these unpopular neighbors to be sited off the east
coast that uses 40 percent of the nation's oil but has only
12 percent of its refining capacity. These, and a dozen
other industries, are serious candidates for islands now
on the drawing boards. Just as to dig or not to dig is,
amazingly, no longer a question, so too these islands
may slip into place, part of the bigger gamble that ignores the coast-in-the-whole.

The trade-off is almost made—a viable coast for the
plunge offshore, for a few more moments of twilight
before the oil lamp goes out, for prolonging the ocean-
sink concept until some version of Black Mayonnaise hits

us in the face, the nostrils or the gut, for expanding industries, for revving up economies. There is some information about what PCBs, Kepone, nuclear plants offshore will do. There are facts about oil in water; there are Louisiana, Stavanger, Santa Barbara. We can anticipate what will come to the east coast, but we don't. Nor do we make a major thrust to fill the fact gap where it gapes, busy as we are inventing ways to get more and more civilization offshore.

To deter any one coastal villain will not stop mounting offshore trouble because no one action is responsible for it. Rather we are caught in a short-term perspective of coastal use, a perspective that puts our heavily pressing immediate needs first, ignores evidence that what we do is inimical to the unalterable requirements of the coast. Our vision is not clear enough to see that further disturbance of the still elusive, still less than understood coastal character in any one of its parts will afflict the indivisible whole and soon end its functioning.

8
WHO'S MINDING THE SHORE?

The Town

The 2500 people of Eastwell,* an exceptionally beautiful Maine seacoast town, are astonished, bewildered and often infuriated by the turn of events that, in the sixties and seventies, have been changing the place they live in. Like many of their counterparts elsewhere on the coast they have been catapulted into trying to decide what their town's future should be, an unaccustomed and to most an unwelcome undertaking. What they decide is specially significant because Maine, with more tidal shoreline than any other state on the Atlantic, and with the last great stretch of wild coast before the United States touches Canada, could lead us to a new coastal

---

*To preserve the citizens' privacy, their names and identities have been changed, as has the town's. In every other respect, this is a factual report of the town's crucial years.

integrity, or follow the trend in the opposite direction.

Eastwell is isolated, relatively untouched (one-tenth of its 24 square miles was developed by 1970), and its splendid coast is enormously desirable. The town's persistent, sometimes balky, resolutely independent and occasionally uproarious effort to mind the shore is at once purely local . . . and universal.

Two peninsulas stretching out into the bay make up this town, providing its extensive coastline and more than the usual Down East difficulty in communication, since it is necessary to travel the entire length of the U to get from one end of town to the other. Eastwell has its fair share of spruce, birch and fragrant bayberry to cover the rock base that sheers off, bald and steep, into deep water. A fifty-acre salt marsh grows in an inlet, lobsters are thick on the offshore rock bottom. Early New England settlement habits persist: small villages clustered around coves with the best anchorages, salt-meadow farms scattered through the uplands. Traditionally, shipowners and captains lived on the higher, more spectacular Neck, crews on the easternmost peninsula opposite, and, at the turn of the century, the first summer people built their porch-trimmed houses out on the Neck, bought local lobsters, fresh fish, and something over half the land, enjoyed placid, multi-generation vacations. Division between the two parts of town was enough to tempt the Neck, twice, to try secession, but by and large the town had a stable, fisherman–summer-people economy. "Comes Labor Day," a storekeeper says, "it's like turning a bucket upside down," and Eastwell is left to its accustomed rural existence.

"If Jesus Christ himself was chairman of the planning board he wouldn't get anywhere," Hank Wood, affable, voluble retired Navy officer, says, somewhat reluctantly approaching his second year in that position. Wood and planning are both unpopular in Eastwell. "I'm still fighting the outsider label," he says. Others attribute his difficulties to being rich enough to live rather well on the Neck, socializing exclusively with summer people and newcomers and being unable to "talk the language"— Maine language, that is.

The unpopularity of planning stems from the towns-people's fierce conviction that it is each man's right to manage his own land as he sees fit. Some say the traditional Yankee independence is the product of hours alone on the sea, lobstering, a demanding one-man task; others attribute it to fear of the unfamiliar. In any case, among people in Eastwell there is an extreme distaste for being told what to do, even by themselves. In the sixties, for example, seven out of every ten new homes in East-well were trailers, the young homeowners being unable to afford anything more. A town ordinance directs that there be at least a quarter-acre lot per trailer. "Almost nobody obeys it," Wood says, "and there's nobody in town who wants to lay down the law to his neighbor's son, setting up housekeeping in the family backyard." Settled in the mid-1600s, the town was incorporated a century later and in the ensuing two hundred years has passed only three laws. The establishment of the planning board is the third.

Alarm sounded in Eastwell when the trickle of new year-round residents became a stream approaching a torrent, people moving out of the fast-growing inland urban center, a forty-minute commute away, moving in from out of state to retire by the sea. Building was going on everywhere, without regard for town resources, and soon the town had a bad case of suburbitis. New land-owners blocked fishermen from access to the waterfront and their livelihood; there were traffic jams on the two narrow mid-peninsula roads; and sewage was sliding off the rock ledges to pollute shellfish in the coves.

"The way of life we know is what most of us want,"

Freeman White, lobsterman, carpenter, boat builder and first selectman, tells a visitor. White, a handsome, blue-eyed outdoor man in his mid-thirties was relaxed after a stormy winter day's lobstering, warming his feet in two pairs of heavy socks ("It's traditional around here instead of slippers"), joking with his children over the supper table, answering three phone calls on town politics, later joining his wife at a meeting of the Historical Society in his living room, the not unusual schedule he had followed for a decade as selectman at $2000 a year. "Everyone loses money on the job," he says. "You have to be interested." He recalls that in the late sixties everything seemed to hit Eastwell at once. "People around here don't pay much attention until there's a crisis. Then they are apprehensive and want to call a halt."

In 1969 apprehension was such that, at the annual town meeting in March, the citizens of Eastwell relinquished their total self-reliance for the first time in history and created a planning board—its goal, "to keep our town just as it is now." White appointed a retired airline executive as chairman ("He had lots of time and *likes* planning"), agreed to the hiring, part time, of a Boston town planner recommended by the State Planning Office, prepared himself for controversy. He hoped the planning board would produce an acceptable plan, having recently realized that a town required one in order to get state and federal aid, but, he says, "I smelled trouble."

It was not long in coming. The new board met once a month, and, eager to get something done, accepted the planner's scheme, modified from one worked out for a Connecticut town. It was a run-of-the-mill, planning-school product, recommending an immediate subdivision ordinance, with a set of severe controls and the ubiquitous study, a three-year comprehensive examination of terrain, population, zoning possibilities. At an

August public hearing on the plan, Selectman White sensed a change in the citizens, an unusual nervousness. "Most who came were against whatever would be proposed. If the ordinance had passed that night, it could never have been carried out."

People in town remember trivia that touched raw nerve endings: the ordinance advised a kind of driveway gravel nonexistent in Maine, revealing the second-hand nature of the plan; the chairman, in a heated moment, referred to the "rubber boot set," emphasizing the town's geographic-social split. Not much of the complicated, two-pronged proposition was absorbed by voters; not many bothered to read the fine print or to puzzle out the technical planning language. What they did absorb was that they were about to forfeit the right to do what they wanted with their own property.

Before Labor Day, more than a year after the planning board had been established, a petition was circulating to abolish it, and a special town meeting was called for September. The opposition organized baby sitters and buses to ensure attendance, and so many voters turned up (more than 600 compared to the usual 200–300) that Eastwell, unprepared for such widespread interest in town affairs, had no place large enough to hold them all. The assembly was adjourned to the next night, when, for the first time ever, the town meeting was held outside its boundaries. The discussion was heated; bitterness, divisiveness poured from the citizens. White had felt the town's pulse correctly. The subdivision ordinance was resoundingly defeated and its creator, the planning board, just as resoundingly ousted.

Post-mortems revealed a split between peninsulas. To those who favored planning, the ouster seemed an emotional binge, a show of ignorance, stubbornness, selfishness. Some blamed it on the "Maine mentality," others on the lack of understanding of what was proposed. "You can't explain red to a blind man," one said. Anti-planners complained that their children couldn't afford to buy land in town any more, attacking the proposal as a scheme to favor the rich "uplanders" on the Neck. "They've taken enough of our freedoms away," a fisherman remarked. "Any more lights on the Neck and even the herring won't come back."

It was not a simplistic division—haves *vs.* have-nots. Not all lobstermen and other Eastwell natives match the prototypical, pure Maine, monosyllabic loner who wants the coast to stay as it has always been, nor are all newcomers devoid of sensitivity regarding the shore. There are some sophisticated, knowledgeable citizens in both groups who are out to make money. Selectman White wants a plan because it will bring needed state and federal funds to town coffers. Jacques Lyons, young carpenter-builder whose French-Canadian heritage is not uncommon Down East, had traveled the United States with his wife and small children looking for a place to settle and had come back to buy an old hotel, remodel it themselves into apartments for schoolteachers and young working people and start a row of tastefully designed low-to-middle-income houses next door. Lyons supports planning: "I plan my own life," he tells a visitor, "why not my town?" He believes the people business is the proper way for the town to grow ("They pay good taxes and cost us very little") and is convinced that growth is inevitable on the shore. A quite different power in town is Sam Hewes, burly general contractor, who travels local roads with his wife on twin motorcycles, was born in Eastwell, where, he says, he has accumulated a million dollars, a hundred acres, and more work than he can get done because his men are "begging to be laid off to collect unemployment—we call it rockin'-chair money— and go hunting and fishing." Because of these labor problems, Hewes is dropping heavy construction and going into paving . . . and development.

With a Portland lawyer as partner, Hewes proposed Eastwell's first development, 134 half-acre lots on some 80 acres of a wild rocky point bordering a pond and bay, in which sand beds would be built to filter an estimated 46,000 gallons of sewage a day. A larger proposal from out-of-town developers on 200 acres of the Neck was for 127 clustered condominiums to cost some $45,000 each. The town was set back on its heels by the double-barreled threat.

"There sits [Eastwell] saying, 'Stop the world. We're not ready for this,' " the local paper commented. But the silent lobstermen and talkative retired executives surprised the world, Maine and perhaps themselves. Pressed by the state for a coastal zone plan, and by the prospect of imminent development, Eastwell took a radical step, becoming one of the first coastal towns in Maine to declare an eighteen-month moratorium on building. It then reestablished a planning board and charged it with preparation of a comprehensive plan. Freeman White said he didn't know if there was anyone alive who could come up with a plan that the town would accept after its first traumatic experience, and it took him three months "to find seven guys dumb enough to participate," as Hank Wood, who was one of them, puts it.

The moratorium was seen as a triumph for nonplanning. True, it held off any more division of land for eighteen months and killed the first two developments; both were rejected by the State Board of Environmental Protection because of the moratorium's existence—and for other reasons. A local fisherman testified that Hewes's pipes would discharge in the immediate neighborhood of thousands of lobster pots, polluting the water where sixty people made their entire living. Other citizens objected because of the effects the pollution would have on a colony of rare seals living near the point, and on the safety of water for swimming. Also in question was the ability of the limited rockbound fresh-water table to supply the proposed hundred or more new households.

Hewes and partners filed a suit against the town, charging the moratorium was illegal, but dropped it after a year, allegedly for lack of funds to continue. Somewhat later, Kennebunk followed the moratorium route, was challenged in court by none other than Levitt and Sons, which was then an ITT subsidiary, poised to condomini-umize the shore. The publicity was such that a lawyer surmises, "ITT may prefer to lose than be saddled with any more of the giant-corporation-versus-small-town image."

By a fluke the Eastwell moratorium was a small and, as it turns out, a short-term victory for those who wanted to keep the town's natural resources in *status quo*. Neither developer could demonstrate the capacity, financial or otherwise, to handle sewage on that nonabsorbent shore without harm to lobsters, seals, people, the fishing indus-try, the tourist business. Long range, the price of tri-umph was surrender of the deep-rooted free-and-easy notion that it is an inalienable right to do what you want with land and water, a right for which Eastwell had fought longer than most east coast communities.

By law the moratorium had to result in a plan. The new planning board, adamant against outside help, hardly knew where to start; apathy was rife. Chairman Wood took a land-use course at the University of Maine ("nuts and bolts aspects") but neither Jacques Lyons, busy with his remodeling project, nor any other mem-bers of the board, had time. There were no meetings in the summer when they were all busy; attendance was sparse thereafter, filled with long silences, Wood recalls, or avoidance of the issue. "I'd start talking about sub-divisions and the next thing you know they're talking about the price of bait or someone's trap line that was cut. If you interrupt, you make a deadly enemy." Discouraged, Wood asks a pertinent question: "How many people in the United States know anything about planning?"

He was soon to find out that in the larger world planning had taken off from the unvarnished days when a man could plan to fish or farm for a living and had become a profession with a language of its own, political know-how and an alphabet soup of bureaucracies. Its nature is compromise. It would inevitably set Eastwell's small part of the coast on its way to a new look. What Wood was really asking was whether anyone in the United States knew what this new look should be, or, even more profoundly (as in the OCS oil question), *if* it should be.

In the sixties the Maine Highway Department had decided to spend some $90,000 of Lady Bird's Beautification of America money on an ancient, narrow, winding Eastwell peninsula road. The state would take sixteen miles by eminent domain, widening the road and installing twenty-two picnic areas, raised curbs, grass shoulders, a big parking area at land's end. Some local entrepreneurs and the selectmen had thought it a good idea, a chance to attract tourists and their money. Freeman White says it never occurred to him that this plan might eventually destroy his own way of life and source of income. Larry Hall, vocal environmentalist, exploded. "When I saw the plan I thought it was crazy. This town needs beautifying with about the same urgency that Da Vinci's Mona Lisa needs a home permanent."

The Highway commissioner stated that he had chosen that particular road "because it was so beautiful." With such reasoning, Hall said, the Great Society's "gung-ho beautification campaign" was not only an offensive waste but a multi-billion-dollar war on privacy that would "change the character of this spectacular part of the Maine coast to a cheap tourist joint." The highway department could have proceeded without town consent, but the protests were so bellicose that the highway men withdrew to beautify elsewhere, and shortly thereafter Eastwell proceeded to the second-hand plan that it soon so vehemently jettisoned. By the early seventies the town had discovered that innocent, honest Yankee independence was no longer able to do without planning and its consequence—growth. The propulsion toward it, now beyond local control, came from the swell of traffic and the sudden passionate interest of the State of Maine in coastal planning.

The State

Maine's remote coast, magnificent in fog or sunshine, its pure water, tasty shellfish and testy natives have become more and more attractive to urban refugees as such resources become increasingly scarce on the Atlantic shore. After textile mills migrated southward, taking the major part of the state's industry with them, the quality of the environment was high, and Maine depended on the magnetism of its character as a prime economic staple, as it once had depended on the mills, the obsolete wooden-ship-building industry, the dwindling fish and lobster catch. But in the early seventies new industries elbowed their way to the coast, competing for space with the people business. Demands for oil refineries, deepwater ports for tankers, energy plants to use cold Maine water, vied with the state's own citizens, who were having an increasingly hard time getting to the shore themselves (with only 34 usable miles of coast in public ownership out of 4000).

Maine's growth rate, one of the fastest in the United

States, jumped from 2.5 per thousand in the sixties to 10.6 per thousand in the seventies, when decades of out-migration switched to in-migration. Seventy percent of the near-million citizens live on the coast. Five million summer people and "travelin' tourists," as an Eastwell gas-station attendant calls them (he boasts of counting ten different license plates in one midseason hour), swarmed into the state for the three temperate months of 1972, 6 million the next year, 12 million are forecast for 1980. At this rate there will be three to four times as many in the year 2000. Planning would have to consider such numbers, either to encourage or discourage them.

Maine had unusual options in the early 1970s because its coast's future was still undecided, and because Maine people are traditionally more tied to and aware of the coast then most. Youngest New England state (it was part of Massachusetts until 1820 and its northeast boundary was settled only in 1910), it had a way to go to catch up to its neighbors in some respects, but in others its very youth and circumstances gave it a unique opportunity. A growing number of Maine men, such as John Coles, editor of the *Maine Times,* believed—and some still do—that the sense of place that inspired poets and painters also gave Maine the chance to be the first post-industrial state. "It's so far behind it's ahead," Coles says.

This radical route, which takes a direction opposite to the trend of the times, was the road not taken. Instead Maine conformed to the patterns of most east coast states, by which standards the acts resulting from its particular coastal fervor appeared unusually advanced, winning accolades in planning circles as the most progressive set of coastal laws anywhere along the Atlantic.

The first consequence of the coast's fast and sudden deterioration was as crisis-oriented and almost as unsophisticated in Augusta, the state capital, as in Eastwell.

Legislators saw the threats to the shore they treasured economically and emotionally, and they set out to "Save the Coast" by passing laws to control despoilers where and when they found them, a direct, action-reaction move of limited vision. It was the specific threat of an aluminum smelting plant and unregulated oil development on the coast, a planner says, that sparked the 1970 Site Location of Development Law, which called for state zoning for any commercial or industrial undertaking over twenty acres to ensure minimal adverse impact on the environment. First of its kind in the United States, it was upheld by the Maine Supreme Court and soon copied by Vermont, Maryland and Florida.

The biggest and newest threat in 1971 appeared to be uncontrolled subdivisions. Eastwell replicas up and down the Maine coast had no ordinances to resist the citizen's God-given right to make his own land-use decisions. The instant American solution in such crises is zoning, a system that appeared in the 1920s, spread quickly across the country, and was welcomed everywhere as the best answer to the land-use problem. Scared legislators in Augusta did not stop to question whether copying this inland device was appropriate to their stretch of uniquely unspoiled coast.

Quite the reverse. Zoning, okayed for decades elsewhere, was happily borrowed by Maine, and its Mandatory Shoreland Zoning and Subdivision Control Law sailed through the legislature, its drawbacks unnoticed in the general huzzahs. Disregarded were long-range implications—a subdivision future for the delicate coast system—and the short-range havoc wrought by a compromise that limited the required zoning to the 250 feet contiguous to the water (as though what happened 300 or 500 or 1000 feet back did not affect the coast), not to mention lack of funds to carry out the law.

Zone by July, 1973, the legislators said to the coastal

communities, or else the state will do it for you.

By the deadline 30 towns had complied with the law, more than 400 had not. The dictum hardly stirred Eastwell's new planning board out of its lethargy. Only Wood recalls getting a notice of the law; there were no accompanying guidelines because there was no agreement as to what they should be.

Temporarily defeated, the state postponed its threatened interference for a year. Some reshuffling and a Ford Foundation grant produced a task force, guidelines, and a model ordinance that follows what Maine says are "sound land use development principles." But even with this added assistance, many towns, regardless of sentiment, had trouble coming up with a plan. Most did not have the detailed maps, technical knowledge or man hours that were called for. Town government couldn't produce more than a mere framework of a plan, and the state planning office, lacking adequate funds of its own, looked to the regional planning commissions (RPC) to help the towns.

Formed in the late fifties and sixties, the eleven Maine RPCs, which the state calls "major community planning assistance agencies for Maine municipalities," have emerged in this era of environmental law-making as overloaded go-betweens. Their job, which includes advice on how to get federal and state money, how to make inventories and identify town resources, is so suspect at both ends—local and state—that concrete achievement is a rarity.

RPCs under many names sprout in various shapes and sizes, fertilized by coastal growth. In Georgia, the Coastal Area Planning and Development Commission unashamedly serves the penultimate word in its title—its goal, "to find the maximum amount of development the available land will accommodate." A giant version is the New England River Basins Commission, with members from seven states, ten federal agencies, six interstate and regional agencies and a professional staff to give technical assistance and act as a planning-information clearinghouse.

"Look at the mess that RPC man made in the town down the road," an Eastwell selectman said. "Why should we join—it's another meeting." But in 1973, when open dumps were forbidden by the state and zoning became a must, Eastwell capitulated, paid its $900 annual dues (fixed by population and tax rolls). "Even if we don't like what they recommend," Freeman White said, "it might be worth the money just to have a spy up there." By then, 235 municipalities had passed zoning laws, 201 were zoned by default by the state with its Imposition Ordinance. Eastwell had made the deadline by three days, was pleased to hear from the state that it had made "one of the better plans," still had no idea how to carry it out.

There was hope that RPC could help with the intractable problem of enforcing the Eastwell plan, involving as it did a multiplication to the nth degree of the difficulties with the comparatively simple but firmly ignored trailer ordinance. "Even the plumbing code isn't enforced," an Eastwell planning board member comments. "One of the selectmen has been pouring raw sewage into the harbor for fifty years. Who wants to stop him? What we need to enforce zoning is a marine sergeant without friends."

The problem was common enough among Maine towns to inspire the Maine Municipal Association to sponsor a thirty-hour training course for people with enforcement responsibilities (although some real-estate operators came too): by spring, 1974, the course had been given in eleven locations to a total of almost two hundred people and needed funds to continue. The technical and enforcement difficulties thus remained largely unsolved and the law a paper tiger. "The best law enforcement in environmental law," a harried state official says, "is a disgruntled neighbor."

Maine's euphoric law-making splurge mixed all the love for the coast, the avid desire for it and ignorance about it into a tangle of legislation, a snarl of overlapping confusion. "Down to the sea in triplicate" is the new necessity, says an observer who considers the instructions of the Ten Commandments easy to follow compared with an application for building on the shore. This does not act as a deterrent to big builders, the ITTs or oil companies like Pittston Fuels, but it does discourage the small contractor-builder who can't afford the necessary paperwork and the once-concerned local citizens who are wearying of the contest. Bureaucratic complexity applies to coast saving, too. In 1970, after a spirited local effort, it was decided to try to restore the Ogunquit dunes, eroding at the rate of 7100 tons annually, to their original 25-foot height. The project had to be reviewed by 8 federal departments, 11 state and area groups. Three years later, when the plan was approved and submitted for review at a local hearing, no one showed up.

Pressure on Maine to make goals, not laws, came from citizen conservation groups of all descriptions, with counterparts in many states. Robert Patterson, president of Maine's Natural Resources Council, which claims to be the principal environment organization in the state, says, "The crucial question for Maine is how the state should develop from now on." Involved in this decision is the state's first environmental intervenor (a voluntary participant in litigation), Horace Hildreth, an energetic Portland lawyer, who, with four or five colleagues, started the Coastal Resources Action Committee (CRAC) in 1968.

CRAC's intention has been to combat the emerging industrial lobby: the paper companies, the developers, the oil interests. It took as its goal one already adopted in Delaware, where, in the pre-OCS oil-dig days when a state could more or less control where oil went, Governor Russell Peterson got a law passed prohibiting any further heavy industry on the already burdened coast. This law, first of its kind, by effectively concentrating the oil business where it was, momentarily protected from further destruction the part of the coast where Du Ponts and other well-heeled citizens live. Hildreth wanted Maine to confine oil to Portland, "already ruined." If Portland was Maine's only oil port, he commented in his bustling law office, it could afford a port authority, pipelines instead of tankers to move oil, better equipment, and, like Delaware, would preserve the quality of life on the rest of the coast. John Coles says CRAC was formed to protect Maine's own Du Ponts, large landowners in Penobscot Bay—Tom Watson, Douglas Dillon, David Rockefeller—from an interloper who had acquired and was about to develop an island there. It was a healthy move, according to Coles: "It shows that citizen intervention can work."

But Maine's laws, however progressive, were conceived within the concept of growth. It would locate industries in one place rather than another, but it would locate them someplace. It would protect parts of its coastal communities but zone other parts for growth. Looking for something for everybody, under pressure from a dozen conflicting claims for coastal use and from citizen groups such as CRAC, confused about the laws and unable to carry them out, the state found itself impelled to try to plan for all four thousand miles of its innumerable islands' and rocky mainland coast.

"To make a plan for compatible and multiple uses of the coastal zone, an undertaking of noble proportions," as an official announcement says, it decided to inventory the coast, classify its lands for suitable future use in zoning-type categories, recommend laws for implementation, and attempt to find out what the public's thoughts on these matters might be. This last, given the citizens' proclivity for silence, was a stupendous undertaking by itself. Starting in 1969, the state hoped to achieve its plan in four years. The going was rougher than expected; an interim report calls the work "a triumph of persistence." After seven years Maine had labored, brought forth— and then withdrawn—a plan for just the middle section of its coast.

The Nation

For almost a decade Congress had been peering through a heavy mist of tradition and limited understanding at coastal troubles such as the nation had never experienced before. The major discoveries about the workings of the coast system—estuaries, salt marshes, sand drift—were just being made, and conflict of use was piling up on the thin edge. Legislators, reaching for ways to deal with the approaching national emergency, found burgeoning in the sixties a new consciousness of land, sea and shore as more than passive utilities. Action for the coast was based on three compatible expressions of what to do with this new idea.

One effort to get national land-use planning, a persistent crusade of Senator Henry Jackson and others, proposed that grants be given to the states to develop their own land-use policies and procedures. A much-noted study of the marine environment, *Our Nation and the Sea*—the Stratton Report—embodied the principles of its progenitor, the Marine Resources Act of 1966. This law makes "accelerated development of the resources of the marine environment" the policy of the United States, and the Stratton Report recommends that it be carried out by moving coastal planning from the towns, where it is influenced by such purely local concerns as the tax base, to the state, which would set up a management system to permit "conscious and informed choices among development alternatives." Two studies of deteriorating estuaries—one in seven volumes—note the connection among coastal elements but blandly assert that the answer lies with state ownership and management of these estuaries, with federal assistance—because 90 percent of the states want it that way.

These traditional views of state sovereignty, variously applied to land and sea, were the foundation stones for the Coastal Zone Management Act of 1972 (CZM), triumphantly proclaimed as the nation's first land-use law.

The act relied on the maxim that the whole is the sum of its parts, a generalization unsuited to the stubbornly unitary coast. Congress further assumed that if state sovereignty was good enough for land, sea and estuaries, it was good enough for the thin edges of the continent which contain something of all three, or at least "this was the best politically we could get," an aide of Senator Hollings comments. There were dissenting voices. Some congressmen agreed with the Sierra Club representative in Washington, known as the number-one east coast honcho, who insisted that coast planning transcends state jurisdiction or judgment and must be a federal responsibility. But Nixon-Ehrlichmann pressure to get a land-use bill first and then attach the coast to it demanded abject compromise. Nixon grudgingly signed the act just before Election Day.

"It is the national policy to preserve, protect and develop the nation's coastal zone," the CZM law announces, and proceeds to divide the shore into state slices—thirteen on the east coast—for planning and management. "The quest for any kind of comprehensive national coastal program has been abandoned," a Wisconsin law professor's study of the CZM Act states, and goes on to describe it as "poorly drafted, deficient in substantive standards, vague on policy." "In other words," an east coast colleague adds, "no good."

Five years later one can hardly ignore the pervasive confusions created by the CZM Act, product of political trade-offs. One can even begin to wonder if the no-good law was, perhaps, planned to fail; certainly the difficulties in implementing it underline its basic flaws.

The act provides for each state to make plans for its share of the coast as it sees fit and for the federal government to spend two dollars for every state dollar spent in this endeavour. When its plan is approved, the state moves to put it into effect—the second CZM phase—for which administrative nightmare (if Maine's troubles with its one coastal zoning law are any indication) it again gets federal funds. A fraction of the budget goes to buy estuarine sanctuaries (the first is Coos Bay, Oregon, the second, part of Sapelo, a Golden Isle). A larger fraction is to run the CZM bureaucracy. Because its location might dictate control, the new office was fought over bitterly for months. Environmentalists and the oil industry, strange bedfellows, wanted it to be in the Department of Interior, where both groups believe they exert influence. Senators Hollings and Magnuson, whose longtime purview is the Senate Commerce Committee, wanted the new undertaking they had shepherded through Congress in the Department of Commerce, where indeed it ended up, under the supervision of water-oriented NOAA (National Oceanic and Atmospheric Administration).

All thirty-four coastal states and territories partake of the federal largesse (except American Samoa, which refreshingly announced it had enough money from another federal source to take care of its planning). The

CZM program is all carrot, no stick, as the states do not have to partake in it. Apparently, however, they find it attractive because it promises that, once the state initiates its plan, all federal monies due it will be spent as the state determines.

Actually, this so-called state planning is no more than the old familiar town planning with a superstructure added. It is the townspeople who must make the inventory on which planning will be based. No state official from Augusta could find the salt marsh in an Eastwell coastal cove, identify the rock ledges which defy development, the possible sites for aquaculture, the beaches, vistas, and dozens of other items needed to make a significant map, a significant plan. And local people are no more high-minded about the coast than anyone else. There is always the calculating local contractor or real-estate operator who would rather sell wetlands than identify them as requiring protection.

Maine met with more than the usual problems because it depended on the regional planning commissions for local contacts, and their deficiencies became quickly and devastatingly apparent. When the state finally achieved its plan for the midsection of its coast, it held citizen meetings in Wiscasset and Ellsworth in May, 1975. Some eighty-five people showed up at each. Comments were mostly negative: "The local government can do it better," the townspeople said, balking at more regional interference. "Coastal zone management is just not that easy to do," the state planner said, resigning after four years of trying. In mid-1975, Governor Longley withdrew the painfully achieved plan for one-third of the coast, declaring that his first loyalty lay with the town governments and that it might take a year or more to get their cooperation. His move was widely interpreted as an invitation to industry.

By early 1977 no east coast state had achieved an

approved plan, and only one state in the nation—Washington—and one city—San Francisco—were in the second CZM stage, putting their plans into effect.

This is disappointing to Robert Knecht, CZM director since its inception, who had believed that "the theory of the Act is that the process will produce the program." A tall, pleasant physicist with a long Department of Commerce career, whose environmental interest was inspired by three terms as mayor of Boulder, Colorado, Knecht's optimism lent an air of excitement to the initial days. The infant CZM was housed in a shiny new government building on the outskirts of Washington, where the staff seemed dwarfed by huge glass walls and miles of bright government-issue carpet. There it assembled starting-line data on the states' coastal thinking. The variety was startling. Five east coast states (all south of New York) had no coastal programs; six were making extensive and time-consuming inventories, hoping then to move on to plans; Delaware, like California and Oregon (and East-well) used a moratorium to give itself time, and Rhode Island arranged to act fast in response to crisis, "the brush-fire mentality." Only a third of the nation's coastal states had more than three years' experience with managing their coasts. Despite the one-room schoolhouse aspects of the job, Knecht dove into it. "I think we can go a long way with what we have," he said.

There was a long way to go and no one knew where they would be when they got there. Through the bureaucratic scramble, regional meetings, state-federal discussion, the serious if not fatal flaw in the law began to emerge. "Massachusetts would have been . . . further into the process of protecting its coastal resources if the CZM Act was never passed," a state senator told a New England CZM conference. "It lets the state drag its feet by the old dodge of waiting for federal action." A Daytona Beach engineer left home at 4 A.M. to tell an

Atlanta conference, "The legislation is a real barrel of worms. Florida . . . needs laws with teeth. This one is a dereliction." "It's evading the issue." "It doesn't say what the national policy is." "Defining national interest is the number-one priority," a Maine planner said. Over and over, such comments bombarded Knecht and his staff.

"There's an urgent need for a national policy," Knecht said during a coffee break in Atlanta. "We need to define the national interest in the coast." Amazingly, this fundamental in the fate of coast, ocean, people, for centuries to come was farmed out to a consulting firm, the MIT Center for Policy Alternatives, for an $80,000 study and, the results being unsatisfactory, for a second $20,000 addition. It has still not been put down officially on paper. Unofficially, it occupies the last chapter of this book.

A CZM-sponsored conference devoted to this one subject of the national interest failed to define it. "Don't be fooled by the words," one environmentalist said, "National interest implies federal override of state initiative." Another, who worked on many CZM contracts, thought the subject "fairly ephemeral." Others said that there is "a clear duty" for the federal government to define the national interest and that to have such a definition, together with the new, stringent dictates of the Environmental Protection Agency, was the only way CZM could work.

But the nation was unable to define what its coasts should be. Special interests rushed in to fill this unnatural vacuum. The Ford White House used presidential initiative to send the CZM Act Amendment to Congress in 1976. It gave the coastal states certain pleasantries— an extra year for planning, upping the federal share of expenses from two-thirds to 80 percent, plus new grants for interstate and regional coordination. Behind this window dressing was the sledgehammer, a $1.2 billion ten-year Coastal Energy Impact Fund, known in certain irreverent circles as the negative impact payoff, "to help States and communities provide public facilities needed to accommodate anticipated inflated populations brought about by offshore drilling operations and other activities." This meant money for roads, schools, hospitals, sewage systems, and thus one definition of the national interest made itself clearly known, providing a strong, billion-dollar incentive to put energy facilities on the coast.

Growth-hungry states lost no time in responding. Massachusetts Governor Michael Dukakis wooed the sixty thousand conferees at the American Petroleum Institute meeting in Houston, hosting a breakfast, two evening receptions and a closed-circuit press conference to announce that a warm welcome awaited oilmen in the Commonwealth, where they would find a fine staging area for the Georges Bank OCS dig. New Jersey's Governor Byrne took a planeload of reporters down the same Houston route, invited to lunch the eleven leaseholders of Baltimore Canyon (seventy miles off Atlantic City), outlined the advantages of locating onshore support facilities in New Jersey. President Carter had suggested the trip, Byrne said, because people might feel that in New Jersey "we're dragging our feet a little bit on our contribution to the energy picture."

The oil industry, which had lost CZM to Commerce, was now overtaking it. Ten thousand copies of a government-produced citizens' guide to coastal management, *Who's Minding the Shore?*,* were locked up in a D.C. warehouse for crucial months while oil-industry representa-

---

*The title of the guide was borrowed from the present chapter while in work.

tives insisted that it carry a disclaimer which implies that it is not objective. The booklet was prepared under contract to CZM by the Natural Resources Defense Council, a public-service organization which uses its staff of lawyers and scientists to compile similar citizens' guides and to prosecute such cases as that concerning PCBs in the Hudson River. The delay and demand for a disclaimer were "the handiwork of oil-connected members of our advisory committee," a CZM executive comments; an NRDC spokesman agrees. This same Advisory Committee for the nation's coastal zone actually held its 1977 spring meeting in Houston so that members could mingle with the sixty thousand oilmen attending their annual Offshore Technology Conference.

Just when the citizens' guide was finally made public, a more troubling publication appeared, a report to Congress from the General Accounting Office (GAO). This was one of several hundred reviews the GAO makes each year in its capacity of legislative watchdog; its purpose is analysis and judgment. *The Coastal Zone Management Program: An Uncertain Future* is smooth, cool, efficient, liberally documented. It takes the CZM program apart, examines its workings, its long catalogue of problems, its apparent inability to make progress. Section headings, like the subtitle, are cheerless: "Resistance to Coastal Zone Management," "Political Support Is Uncertain," "Federal Participation: A Major Problem." States with coastal legislation are now fighting to repeal the act, the report notes; no coastal laws have been strengthened; and the climate for environmental programs involving government has become "much harsher" since the act was passed.

By innuendo the GAO appears to be giving CZM the ax, or at least a somber warning. "Yes, it's pessimistic," says William Martino, GAO assistant director and NOAA-watcher. "Personally I like the program—it's all we've got for the coast—but its prospects are uncertain because of vested interests—oil companies, developers, not necessarily bad, but powerful pressure groups." Martino had a hand in the minor aspects of the 1976 Amendment but, he says, the billion-dollar negative impact fund was recommended by the Congressional Research Service. "We were told to stay away from that."

Vested interests do indeed make the outlook for CZM pessimistic. By the fall of 1977, an ambitious national ocean policy study undertaken by Commerce threatened plans for a White House Conference on the coast: "[It] could eclipse our efforts to draw additional national attention to the coastal portion of this larger set of problems" Knecht says. If the pursuit of OCS oil and vast development are to define the national interest in the coast, federal policy and CZM activity, however important, will only be in the way. Hence the proffered bribe of negative impact money, and, if this bribe doesn't work, protesting states can go to court, as California has already done to challenge Department of Interior decisions about oil leases.

Or, the GAO report suggests, the states can drop out of the CZM program. And that will be the end of the nation's first effort to plan for its coasts.

The political courage to exact a forthright coastal law was not to be found in 1972. There was no declaration, then, of the importance of the coast to the nation. It was impossible for the states on any coast to deal with the overall coast problem. It was impossible on the east coast for CZM to weld thirteen separate sets of coastal plans into any purposeful national action. It was a leaderless time as it was a time of countless contending leaders, each vying for what his followers wanted from a particular part of the coast and thus destroying the chances of getting it. It was a time that by default advances the nation's encounter with its true interest in the coast.

9
FIRST STEPS

Everywhere along the coast is evidence that we are headed, nonstop, toward developing each inch of it. It would be naïve to suppose that without preparation we would suddenly reverse the determined march of what many believe to be progress, or find instantly convincing reasons to change our ways. Big money and big government listen to a different drummer from the one Thoreau had in mind, and furthermore the voice of the hurting shore is still low key. Even the most dedicated romantic would have a hard time believing that the natural function of the coast will be so generally understood and respected next week or next year that its requirements will dominate the coming drama of the thin edge.

But two potent forces are likely to be overlooked in conceptualizing what is next for the coast, one being too ordinary to be seriously considered, the other being almost unknown and therefore unnoticed. The first is the profound emotion aroused in most people by the grandeur of the coast, love of the kind that inspires fierce protection of its object, partly selfish—"I want it to be there for me"—partly an appreciation of its universality. Tied to this love is the new terrible fear that the coast, refusing now to support various kinds of life, already shut off from view, impossible to reach, its ocean poisoning fish, suffocating birds in oil and burdened with refuse in its very depths, may reject us too. Many generations before us have experienced this love; none has experienced this fear.

Everyday emotions—love and fear—are odd entries in the titanic struggle roaring along the Atlantic shore, islands, and Outer Continental Shelf. But already such feelings have moved some men and some communities around the globe to act to protect the coast. The concrete results that have been achieved are important preliminaries to a radical change of view. They are often incomplete, imperfect, even injudicious, but their very existence assures us that we have the capacity for such enterprise. We can think the unthinkable as Christopher Stone's *Should Trees Have Standing?* argues. We retain the capacity for adventure that we were born with.

And we are at least adventurous enough to widen our options. Salt marshes disappear too fast, but a marine scientist does transplant *Spartina* seedlings onto dredged sand flats, starting new tidal marshes. There are no longer enough fish to go around, but the Russians now harvest super-abundant Antarctic krill, believing we could learn to ingest the unappetizing "ocean paste" as a source of protein. Mariculture has become big business, growing huge crops of mussels and kelp, and making use of the homing instinct of salmon through transporting the young to hospitable waters whence great numbers of adult fish return (Japan recently released 1.3 billion from one hatchery alone). Planners look for new uses for urban waterfronts now that the decline of cargo and passenger shipping leaves them almost empty. New possibilities for disposing of waste vie for funds with new ways to create energy. Some coast towns limit their populations; some private organizations buy miles of coast to try to keep it "unspoiled"; some government powers begin to recognize that deterioration of one part of the shore or another adds up to a massive national dilemma.

Ideas and actions that appear scattered and unconnected, when seen together constitute hope for a restored coast, for a renewed chance to enjoy a clean, productive shore and even for learning to live by its dictates, for moving from conquest to attentive stewardship. As ideas grow stronger and actions more numerous, their significance mounts. Our common hopes for the coast are woven into many patterns, from the work of a master seaside gardener to the emerging application of our judicial system's wisdom to the quality of the environment.

An abundant shoreline edges Jamaica Bay, abutting New York's Kennedy Airport. This area, so rich in natural treasures that it is known as the ecological Fort Knox, is the almost single-handed achievement of Herbert Johnson, a New York City Parks Department man installed there in the early fifties by Parks Commissioner Robert Moses and left to his own devices. When he retired in the seventies, Johnson had created a compelling coastal landscape of such variety that it attracts more than three hundred bird species. "There wasn't even a robin around when we started," Johnson tells a visitor.

The potential of the nine-thousand-acre marsh had encouraged Moses to trade sand from the bay to the Transit Authority in exchange for dikes to impound fresh water in two ponds and for organically rich sludge from a nearby sewage plant to be piped onto an island as fertilizer. Johnson planted grasses and berry-bearing plants, made the island an Eden for nesting shorebirds. Elsewhere in the marsh are groves of Russian olive trees and Japanese black pines he started from seed, thickets of beach plum and *Rosa rugosa* grown from cuttings collected on his travels, fields of wheat, oats, rye. The expertise required for this transformation developed when Johnson was an apprentice to his father, a gardener on the Rockefeller Tarrytown estate, and later a horticulturist testing grasses for Parks Department golf courses. This he combined with self-taught ornithology and the clear determination to restore the marsh to its wilderness functions.

Growing conditions were poor and there was no regular budget, but in five years there were wild ducks and geese in the ponds, shorebirds, land birds, settling in. By 1960 snowy egrets had returned to build a colony; the next year, glossy ibis, gone from the bay for ninety years, rewarded Herbert Johnson by producing what is now a faithful one hundred pairs. School children gaze at the moving scene in awe, obey the sign "NESTING AREA—KEEP OUT," learn migration patterns well enough to expect traveling terns, plovers and the rest on schedule, an unusual educational bonus for these city kids.

Twenty years after the Transit Authority had built the ponds, it proposed to fill in the east half of the bay to add runways for Kennedy Airport, thus destroying at least part of the refuge it had had a hand in creating and increasing bay pollution by inhibiting tidal flushing action. Even if the refuge is spared, the National Academy of Science said that the birds will be hazardous to planes flying nearby and will have to be eliminated. But the more immediate threat is the new management of the refuge. It is now part of the National Park Service's struggling Gateway undertaking, under which aegis the once carefully tended shore has become overgrown, unkempt and dangerously uncared for.

"They're not maintaining it properly," Johnson says. "I thought the federal government would keep it up. They can run the Grand Canyon; you'd think they could run this." He has discovered that institutions without a strong mandate to protect what they control have a hard time of it.

On the other hand a by-product of Johnson's labors was the neighborhood's devotion to this bit of coast. Bird watchers who had grown up with the refuge protested its demise so vociferously that the besieged Gateway management budgeted a "gardener with the proper background" and promised that NPS would restore and maintain the preserve. That the Park Service will by its nature be unable to keep that promise is less important than the existence, for a time, of the tangible result of one man's feeling about a marsh.

The same instinct in the cities around the shores of San Francisco Bay produced the first large-scale coastal rescue operation in the United States and, some say, to this day the most successful of those that have followed.

There was nothing subtle about what was happening to the bay in mid-twentieth century. It was being "reclaimed" (the word itself is a land-conquest holdover) at the rate of 2300 acres a year, twice as fast as a century before. As people poured into the bay area, the demand for filled land zoomed. In the fifties, a Corps of Engineers study forecast that the bay would be little more than a river by the year 2000. Citizens who relied on the sight of the sparkling blue expanse beneath their windows, particularly Mrs. Clark Kerr, wife of the University of California's president, formed a Save the Bay movement. She persuaded a university planner to make a second bay study, and more and more people and such institutions as the Washington-based Conservation Foundation joined in the bay-saving effort. In 1965 the legislature, convinced that a scarce resource was at stake, appointed an interim Bay Conservation and Development Commission (BCDC) to regulate and control waters, marshes and shorelines. In 1969, BCDC became a permanent part of the bay scene.

Its first goal was its *raison d'être*, to stop the shrinking of the bay. Instead of the 2300 acres filled each year pre-BCDC, five acres were filled in 1972, four of them to further the other goals BCDC had defined for itself.

One of these was to increase public access to the bay. For this BCDC was given jurisdiction by the state over a hundred-foot-wide shoreline strip so that it could insist on the public's right of access, some form of which was included in all seventy-one permits of its first five years. Another goal was to improve the aesthetic value of the shoreline by requiring every structure to fit a plan that separates industrial, residential and recreational use of the bay shore, and to pass a Design Review Board. Twenty-two percent of the proposals were approved as submitted; 66 percent required change; 11 percent did not make it at all.

For knowing what it wanted to do, finding support for its goals, fitting its structure to achieve them, BCDC has been admired, analyzed, upheld as a model. Its great advantage was the bay itself, contracting fast enough to stir the protective instinct in those who watched it shrink each day of their lives. The challenge was met; the bay is as secure as it can be in the new circumstances of the seventies.

Success made planning infectious. Many San Franciscans wanted to extend protection to the oceanfront, and people throughout California—spending much of their lives outdoors or on freeways passing the shore—were acutely aware of the coast's rapid deterioration. For years, control of the more than 1000-mile shoreline had been fragmented under the jurisdiction of 15 counties, 45 cities, 42 state agencies and 70 federal agencies. Concern was diffuse and ineffective; proposed coastal protection bills got nowhere during three successive state administrations. The success of BCDC inspired and fueled a movement on behalf of the coast, and its staff and commissioners helped weld conservation groups into the California Coastal Alliance in 1971. But this organization, too, failed to get a bill past the tough legislative opposition, proponents of growth on the coast.

Angry, frustrated and determined, the Alliance persuaded half a million citizens to supply the required signatures that would put coastal legislation directly to the voters by referendum, at the same time building up a coastal constituency. A million dollars or more opposed the effort, contributed by the California Chamber of Commerce, the Building and Construction Trades Council, Standard Oil of California, General Electric. But on Election Day, 1972, 55 percent of the voters cast their ballots for the Coastal Zone Conservation Act, a landmark action better known as Proposition 20.

"The people of the State of California hereby find and declare . . . that it is the policy of the state to preserve, protect and where possible to restore the resources of the coastal zone for the enjoyment of the current and succeeding generations . . ." Proposition 20 said. It provided one state and six regional commissions for a four-year term, a total of 84 commissioners serving without pay to control by permit any development in the coastal zone (1000 yards inland, 3 miles seaward) while it was making a permanent plan. BCDC's chairman and its director migrated over to take charge.

It was an extraordinary move, unique in the nation, coupling permission and planning in an agency that was the creature neither of the state nor of the localities but established directly by the voters and dedicated to the coastal imperative. More than half California's citizens wanted the commission to preserve, protect and restore; it was, in effect, a public trust. "In some ways it follows the Talmudic injunction: let us do and then let us talk," Melvin Mogulof says in *Saving the Coast*.

What it did had to be done on the cheap, as Mogulof says, there being no money to acquire sites or compensate people for losses. Of the more than 5000 requests for permits filed per year, the commissions granted over 90 percent, either because of political realism (the day of reckoning was very much in everyone's mind) or because making these "real world" decisions while simultaneously making a plan clarified issues the plan would have to deal with. "We're getting flak from both sides," a commissioner says, "so we must be doing it just about right." Nonetheless, many believe the coast's condition warranted more than a middle-of-the-road consensus.

Thus a hold-everything moratorium was sacrificed to expediency so as to assuage the opposition and counter the charge that a moratorium would so reduce land values and local taxes that it would bring about a severe economic depression in every one of the fifteen coastal counties. Ellen Stern Harris, the commission vice-chairperson who had helped get Proposition 20 under way, says that without the moratorium "instead of saving the coast Proposition 20 seems to have all too frequently legitimized its destruction." But the commission's director found the short-range ad hoc judgments less problematical than ability to plan for the future. "The big trick is trying to see ahead," he comments. "What is the carrying capacity of the coast?"

An attempt to define that capacity was included in the plan. "The coast should not be treated as ordinary real estate," it says, "but as a unique place where conservation and special kinds of development should have priority." It spells out protection, gives priorities to coast-dependent development such as ports, stringently controls biologically sensitive areas (wetlands, estuaries), provides for increased access and directs development and energy facilities to sites where they will do the least harm. Local governments—town or county—are required to adopt regulations set forth in the plan under the eyes of the regional commissions, which will then cease to exist (remarkable in itself in this age of self-perpetuating bureaucracy). The permanent statewide citizens commission to oversee coastal activity has the power to enforce regulations; its teeth are stiff fines for violations.

In the summer of '76, the bill to make the coast commission permanent was close to perishing in the legislature. In a last-minute rescue attempt, young Governor Brown had to do battle with ex-Governor Brown, his father and chairman of the major anti-bill organization, and with development-minded senators. He pulled a compromise out of Sacramento's smog, and the bill, effective January, 1977, became the most comprehensive state coastal law in the country, providing a permanent structure to protect two-thirds of the nation's west coast.

It moved coast decisions out of local hands. The California counterparts of Eastwell's planning board would now operate under state direction . . . to a degree. The law is "far less heroic than the preliminary version," one analyst says, and there is general agreement that, although it retains its protective thrust, it lost some of the clarity that illuminated BCDC by the compromise required to get the law passed. As Proposition 20 veered away from the moratorium, so the law takes economics and politics into account in deciding on the carrying capacity of the coast, and does not stipulate the ultimate no-growth objective.

But it is a step ahead for the entire country that California has established a new concept of minding the shore and that millions of its citizens recognized the requirement of special attention for the easily disrupted coast system and voted to make it subject to special state controls.

Long range this step is particularly important because it is a small step taken by a state that wanted to make a major stride. California, like other states, is held back by the difficulties in taking rights away from landowners, no matter how important such a move might be for the public good or even survival. Throughout our history, as land once held for the general interest has shrunk, we have designed intricate statutes to protect individual property rights, made them sacred unto the holder of the land, and legalized the philosophy that once inspired a memorable verse:

> *The law detains both man and woman*
> *Who steal the goose from off the common*
> *But lets the greater felon loose*
> *Who steals the common from the goose*

The bundle of property rights—the right of use, of disposition and of capital gain—is lighter now than when the magnificent rocky promontories and golden beaches of the Pacific coast were settled in the nineteenth century. The state's new restrictions make it even lighter, but this had to be accomplished with infinite care. Just as in Maine the tiny step of imposing on towns a 250-foot mandatory shoreline zoning was as extensive a move as seemed politically and legally feasible, so in California, legislators had to keep their political viability intact and stay within the existing legal framework.

Every state has certain powers over its lands.. With the power of eminent domain, a state can take land for public use and pay fair market value for it, hardly a practical solution for such as California's immense coastline. With its police power to protect the health, safety and welfare of citizens, a negative power that says what you cannot do on your property, each state can legislate in the public interest. If it denies you freedom of use by this police power, you are not entitled to compensation but are assured due process of law. Between eminent domain and police power is the power of accommodation that might restrict the owner to reasonable beneficial use and provide for tax benefits or payment on a lesser scale than "just compensation." State controls can be and often are delegated to smaller units—towns, counties, regions—where the decision is even more difficult than for the state, the view being more parochial, the attack more personal. Georgia's still-wild Golden Isle, Ossabaw, had its destiny continued to be delegated to the local tax assessor, would be as certainly headed for profitable development as the island of Martha's Vineyard, Massachusetts, which, guided by a local commission, has already arrived there, sacrificing its natural assets in the process.

No one is quite sure how to define state powers. "The general rule," Justice Holmes said, "is that while the property may be regulated to a certain extent, if regulation goes too far it will be considered as a taking." How far is too far? More than fifty years later, the question remains unanswered with any certainty, Frank Grad's *Treatise on Environmental Law* tells us. A substantial and compulsory reduction of what a property owner considers the cash value of his land not only raises questions of constitutional law but also of fairness, Grad says. In a society that has encouraged land speculation for so long, we must now explore whether it is fair to ask the owner of ecologically significant land—a dune, a marsh—to carry the costs of preserving it rather than to bill the public for whom such assets are to be protected.

The power to protect exists in the states although willingness to use it and to pay the bill for protection is not widely apparent, not yet. But there is reason to believe that we may change our concept of responsibility for the coast, that first steps will lead to next steps.

It has happened before. The concepts that have gotten us into our present fix are light years removed from the ideas of our predecessors on these shores, the Indians, who believed themselves the land's stewards rather than its owners. Across the Atlantic, the Scandinavian countries have become technologically advanced without serious destruction of their natural resources; so, to a degree, has Great Britain. In the tiny Shetland Islands off Scotland, distinguished by a stark landscape and an existence centered around the croft and the sea, the fiercely independent citizens, without the benefit of sophisticated planning, have enacted tight controls to govern the impending invasion of the oil business.

Had you grown up in Sweden before its recent coastal laws were passed, you would have assumed as a matter of course that you could get to and use any of the 8500 miles of mainland coast, the 100,000 lakes, the rivers—*Allemansrätt,* as the custom is called. When the coast started to become crowded, the custom was formalized into law. In the late sixties, according to Richard Hildreth's 1975 study of Sweden's coastal controls, use of the coast did not seem sufficiently regulated to satisfy the water-loving Swedes, and they set about making their

1971 National Physical Plan. The plan dictates that three coastal areas will have no industry, three will have no new heavy industry, no new industrial areas will be opened on the west coast, and there will be no expansion of leisure housing on most parts of the southern coast and in the Stockholm area. Sweden has no constitutional requirements that it must pay for taking private property for public use, and after certain reforms and the National Plan, it was able to include in protected zones without compensation more than 12,000 miles of coast and inland shore. "The impact," Hildreth says in admiration, "appears to be spectacular."

A generation or two ago it was believed that the Constitution prevented the United States from taking national action to protect its coast as Sweden has done. Today that belief no longer prevails. Federal authority under the Constitution has been so extended by the Supreme Court, according to Columbia law professor Albert Rosenthal, that it can undertake "virtually any conceivable measure reasonably intended to protect the environment." The tradition of states' rights in respect to land regulation is potent and Congress has historically been reluctant to challenge it, but that has been the legislators' decision, not the Constitution's nor the courts'. "The federal government clearly has the authority to regulate the use of private property," Grad states, explaining that this authority flows from powers conferred on the federal government by the Constitution—the power to tax and spend, the power to make treaties, and, perhaps most applicable, the power to control interstate commerce, the basis, for example, of federal laws concerning air and water pollution, which both cross state lines.

The first national land-use law, the Coastal Zone Management Act, delegated planning to the states, financed their efforts, then came dangerously close to perverting the law to serve special rather than public interests. The billion-dollar negative impact purse that CZM offered the states in 1976 eases the way to OCS wells for oil companies now on the ready.

But, as the judicial branch continues to interpret constitutional power, our culture moves toward the notion that the nation's rights may supplant those of the individual, the town, even the state, and that owning land cannot guarantee the untrammeled freedom it once did. This is not to suggest that shore-owning Americans are joyfully relinquishing their expectations of coast-based fortunes. On the contrary, as regulations become more restrictive and the rights bundle ever lighter, there are more appeals for justice, for the inalienable right to due process of law.

"Courts alone cannot and will not do the job that is needed," Joseph Sax, Michigan law professor says, "but courts can help open doors to a far more limber governmental process." The moratorium, for example—an enforceable public right to delay action when more knowledge is needed—is central to emerging environmental law, part of the developing concept of the public trust. The court's great strength derives from its objectivity, its obligatory non-involvement in the issue. Sax tells of a judge in an environment case who had to look up the word *ecology* but could still ask the all-important questions of balance.

In courtrooms across the country, judicial decisions support the idea of coastal use for the public weal. In Florida, a judge ordered a developer to restore the mangrove swamp, dredged without a permit to become a fifty-acre trailer park, serving notice that "despoiling natural resources for private gain is no longer a natural right." Four other builders were ordered to return ripped-up segments of Key Largo and other islands to their pristine state. In Washington the state supreme court restricts automobile traffic on ocean beaches, "a gradual withdrawal," the judge says, "from the concept of ocean beaches as highways." In California the courts are known to have been "remarkably responsive" to the goals of Proposition 20 and the law that followed it.

An exceptionally dramatic interpretation of our new environmental awareness, a step that might be considered a quantum leap, was Supreme Court Justice Douglas's minority opinion answering *Should Trees Have Standing?* with a resounding yes (following an idea proposed to Douglas by law professor Christopher Stone in the hope that the environment's interests would be taken into account "in more procedural ways," Stone said). "Contemporary public concern for protecting nature's ecological equilibrium should lead to conferral of standing upon environmental objects to sue for their own preservation," Douglas said, arguing that since a ship or a corporation is an "acceptable adversary," why not the valleys, meadows, lakes, rivers, groves of trees? If lawsuits could be brought on behalf of such natural entities against those who damage them—specific polluters, for example—and if the courts continue to insist on justice for such as mangrove swamps and sandy beaches, this judicial opinion could lead to a solid advance.

A start in this radical direction came about with unexpected mildness in an unheralded, noncontroversial 1970 law—the National Environmental Policy Act (NEPA)—that directs all federal agencies to take into account the environmental consequences of their proposed actions. The Act set up a Council on Environmental Quality to advise the President and, almost as an afterthought, required a detailed statement of environmental impact for each action using any U.S. funds. Soon there was an Environmental Protection Agency (EPA) to carry out the law and a chaotic landslide of impact state- ments (EIS)—about one thousand a year. "All [the law] said in essence was 'Look before you leap,' " Gladwin Hill reported in the *New York Times*. "But that in itself was almost a revolutionary notion."

The impact statements are expensive—it cost some $9 million to compile the EIS on the Alaskan pipeline— difficult to put together and generally long-winded, so much so that they are often not carefully read except by zealous environment watchers, who have brought some seven hundred suits to court in five years; in those suits that were completed, two-thirds of the complaints were found justified. "If Congress had known what it was doing," one of its members says, "it would not have passed the law."

This law, requiring that we look before we leap and make full disclosure of the anticipated results of spend- ing the public's money, brought about a recent spectacu- lar holding action: the U.S. District Court's decision to cancel a $1.1 billion federal sale of oil leases in the Balti- more Canyon, the first time such a sale has ever been voided in the courts. With the assistance of the Natural Resources Defense Council, the suit challenged the Sec- retary of Interior's EIS on these oil-lease sales, and the judge's 132-page decision found that the Secretary had violated "the letter and the spirit of NEPA," that his failure to grapple with specifics bordered on irresponsi- bility, that the decision to lease had been a foregone conclusion, making the EIS a "charade," and that "the public's rights and equities are paramount and must pre- vail."

A new EIS and a well-planned appeal may bring the oil rigs into place off New Jersey after all. The U.S. Court of Appeals reversed the decision within months. But front-page headlines announcing more and more frequent oil-tanker accidents, the uncertainty surrounding the effects of OCS oil on coastal life, the mounting fears contribute to the current strongly voiced concern. Environmentalists are more than a little encouraged by knowing that you *can* beat City Hall, at least in plans for the Baltimore Canyon for at least a season of 1977.

The brief Baltimore Canyon victory is one of hundreds of actions which indicate that we are, amazingly and almost without knowing it, taking an evolutionary step. We are becoming a people who will battle to keep the coast intact. The challenge now is to funnel emotions toward making the change in our attitude one that is clearly enunciated and firmly fixed in the life pattern of our society, and to do it with the utmost urgency.

In 1977 a flurry of academic analyses of the challenge gave us a measure of current thinking among those who deal in models, scenarios, alternatives. MIT's Michael Baram reviews laws regulating the siting of facilities, "the heart of coastal zone management," believes the challenge is "the management of growth to meet multiple objectives." Robert Ditton and two Texas A&M colleagues say that the challenge is "regulating man's use of the ecosystem and attempting to merge the bureaucracy and the market to achieve this task," conclude that the problem is one of human systems and the goal—"to define and preserve the nation's coastal-carrying capacity."

The east coast presents the challenge in extreme form. High-powered interests vie for its control and for the profits such control produces. It is subject to the laws of thirteen separate states, each with its own requirements. It becomes more urban, more polluted, more crowded every day, is every day under more pressure to yield its natural resources to exploitation for the growth society. It is immensely more complicated than any other coast and thus requires far more clarity of vision.

The admirable gardener of Jamaica Bay had instincts that are now widely shared, are much in the public mind and have been translated into action in great variety. Whether or not such action will be found effective or will be reformed or replaced, the very fact of its existence underlines the importance the thin edge is assuming in contemporary life. It is remarkable that the coast should have been the subject of the nation's first, and to date only, land-use law. In retrospect the Coastal Zone Management Act fell far short of the need. Expressing no national purpose, it followed the traditional pattern by turning the problem back to the states, burrowing into a multiplying bureaucracy and becoming one more example of what Justice Douglas succinctly describes as sending the goats to watch the cabbage. Nevertheless the law exists, and that alone moved our evolution ahead.

The citizens of Eastwell showed us the intractable trouble one small coastal community can have in coming to a decision about the use of its land, a trouble testifying to the parochial perspective that exists in every coastal town in the nation. The same perspective, many times multiplied, caused Californians to react against such parochialism, take such decisions out of local hands and present them to a state body appointed for that purpose —even though that step was hampered by conflicting interests which obscured its goals and deprived it of a sharp direction. California has not achieved the clarity of knowing what the coast must be that gave the people of the tiny Shetland Islands the towering authority to restrict giant oil companies, the same absolute clarity that accounts for the much-admired BCDC success.

Laws are evolving to encourage that clarity, even though new interpretations of the Constitution raise some difficult issues regarding the transition from the era of conquest of the coast to the era of its protection. Economics, fairness, the rights hitherto inherent in the ownership of land are now in question, requiring us to resolve who should pay the costs of protection, whose rights are paramount as between property owner and public, between river and despoiler, fish and fisherman. We must deliberate whether the chance to look before we leap, by EIS or moratorium, is justified if it causes economic grief, and we have to know what our goal is.

There is very little time. Cancerous growth of what we put on land and in the water spreads mercilessly fast, affecting every form of life in ways we could not have imagined yesterday. Quite certainly there is not enough time to understand the coastal system completely before deciding how this generation—and those that will follow for many centuries—want to use the coast. But first steps, here and around the world, suggest priorities, demonstrate that the overriding concern must be the coastal connection.

Our list of endangered species is archaic. We can't preserve a single coastal species outside of an aquarium or zoo, any more than we can save one island, one stretch of wild shore. It's the endangered coastal ecosystem in its entirety that is at stake.

For this indivisible whole we need an indivisible solution.

10
THE LEAP

The thin edge writes its own scenario. It defines the national necessity:

TO RESTORE THE COAST TO ITS OPTIMUM
NATURAL FUNCTION AND KEEP IT THAT WAY.

Only the nation can remove the terrible threat of an overburdened coast; only the nation can establish and carry out a program for the protection of all Americans. Discovery of the linkage between all coast systems is the key to what we do about the clear and present danger. It delivers a mandate to restructure the piecemeal efforts on the coast of the past years and weld them into a single governing entity.

No more information is necessary to substantiate present danger and coming cataclysm. Meager as our present knowledge is of the thousands of coast systems, it is enough to support the radical requirement of a single national program. The exponential timetable of mounting disaster does not allow leisure for debate, deepened understanding and a carefully-mounted propaganda campaign. The uncommon nature and condition of the shore require us to move fast, and the most convincing persuader is each reeking beach, each contaminated clam, each sewage-clogged cove, the latest oil spill.

To be responsible for the coast, the nation must reverse the flow of power sent out to the states, most recently by the CZM, which encouraged each to take care of its own coastline and which most states passed on to counties, cities, towns. To say that the authority must now be gathered to the nation is not to deprecate the role of the states in our federal system but to recognize that an indivisible solution is required. On the east coast, as has been seen, the goal cannot be achieved through thirteen separate sets of intentions.

There are existing means to achieve a national program. Congress can endow a coastal body with the tools, use the existing CZM budget to put it to work immediately. Congress can authorize a moratorium on all new activity on the coast to allow time for deciding how coast usage can best be controlled by the exercise of the spectrum of available powers. And there are indications that the courts are moving to uphold the necessary lightening of the bundle of coast rights. Amending the CZM law to create the needed structure would save time and money; it would also have the advantage, real and symbolic, of modernizing the thinking that prevailed when it was devised. Or a new law could replace the old. The fastest way that a unit to restore the coast can be created is the best way.

The purpose is clear. The coast functioned in all its magnificence with no help from man before the present

burdens were imposed on it. Presumably it could do so
again. But it now requires help. It is not enough to be
defensive, measuring how much is too much for the
coast, trying to guess its "carrying capacity," not enough
to stay with the environmental impact statement, important as that has been in some circumstances, since it
implies that a little impact is perhaps acceptable, or that
an oil well in one offshore location is better than another.
The new coast program must instead be aggressive, en-
dorsing the only move man can make to restore the
shore, which is to leave it determinedly alone.

If any activity, existing or proposed, interferes with
restoring or maintaining a natural function as minuscule
as the transparent little ghost shrimp, as majestic as a
torrential river, it should have no place on the coast. The
sole limits are existing physical conditions—New York
City is somewhat firmly fixed—and such necessary activities as can't function anywhere else: fishing fleets, fish-
processing plants, docks, wharves, navy, coast guard, an
oceanographic institute. The goal is limited too by irrev-
ocable changes already made, the extinction of some
species, the permanently altered ocean solution. There
is no hope, for instance, that the east coast can regain the
abundance of swarming fish and towering trees found by
the first explorers, or the pure clear water and virgin
sand. But nothing else is immutable, even our attitude
toward the coast.

That attitude is already changing, as we have seen.
But leaving the coast alone to recover still entails a major
cultural value switch. We will have to revise in fast order
long-held ideas about the coast. It cannot continue to be
a source of indiscriminate economic gain; the people
business will have to move off the coast, as will industry.
We will have to find other methods for disposing of
chemicals and wastes, substitutes for dredged sand,

other places for the whole catalogue of accustomed profitable coastal uses. What must remain on the coast will be stringently regulated by the national body to impinge as little as possible on the coastal function. Growth, if growth continues, will have to move inland within the next year or two, to ground not so delicately balanced as the coast or so tightly interconnected.

For this to happen we will have to give up something, and that something is likely to be pleasurable or profitable or both. People responsible for the new coast program will have to invent the skill to deal with the inevitable howls of protest from some exceedingly powerful antagonists. Change is painful, particularly change involving the loss of an accustomed freedom. But pain and fury, outcry and in-fighting are part of the awesome duty of coastal rescue. Optimum natural function cannot come about without it, and the factor of indivisibility knocks out familiar accommodation by trade-off and compromise. A weakened link weakens the whole interlocking system.

New abilities to turn away wrath will spring directly from the coast condition. The eagerness to keep the coast viable for love of it or fear of it is evident in the recent record wherever people have recognized the necessity to take care of the coast, where they have been trying to regulate and have consistently and terribly failed, unable to do more than fasten on the needs of one small segment or one specific power complex. Given a national goal for the coast, this eagerness might intensify, spurred by the anticipation of resolving conflicts such as those that tore the Eastwell town meeting to shreds, instigated sludge-centered lawsuits between Long Island towns, blocked a safe future for Georgia's pristine Golden Isle. Lagging enforcement of regulations, requests to call in the marines, could advance to determination, spurred by a solid hope if not conviction that the goal for the entire coast has an excellent chance to succeed.

There could be real satisfactions on the way to success. Moving inland would be the act of a patriot, serving a national purpose as important as marching off to war. There are aesthetic delights to be enjoyed in woodland lakes, expansive fields, mountains, long valley views. There might be very little living on the shore, but there could be instead room for a U.S. version of *Allemansrätt,* so that people could get to the sea easily without hurting any coastal system. It might mean a longer walk, a shorter stay, but it might also mean free access to a safe and sparkling place.

The most difficult aspect of this incredibly difficult endeavor is the predestined encounter with the energy industry in its assembled power. To legislate the oil business off the Atlantic coast, in all but its minimum essential pipelines, to forbid OCS oil wells, to prohibit dangerous tankers, to insist on inland refineries is the major battle in the war for a viable coast. Add the nuclear-energy plants on rivers, the plants-to-be on artificial offshore islands, and the defenders of the coast encounter an adversary of giant proportions.

But to pretend to try for the goal—restoring the coast to its optimum natural function and keeping it that way —would be farcical if this battle were not joined and won. Whatever the price to the nation, in effort, invention, money, it is a clear imperative. We know too much about the effects willingly to raise the temperature of the ocean by one more fraction of a degree, to change its contents by one more oil spill. Short-term gain cannot but lead to long-term disaster.

The thin edge needs a champion on the highest level to give it at least equal importance with the energy business of the nation, recently recognized as deserving its own United States Department with a $10.6 billion 1978 budget. If energy is not to dominate decisions about our existence, we also require a Department of the Environment, to provide balanced judgment, to put forward an equally pressing moral issue, an equally challenging war. This department is needed by many parts of the country now under extreme stress, has been urged in many quarters for several years. The dangerously deteriorating coast today provides an immediate insistent reason for the President and Congress, if they recognize the significance of the indivisible coast, to create such a department.

With this champion, the case of the coast can be strongly put to those who control energy as well as to the established representatives of Commerce, Housing, Transportation and Interior, all of which use the coast for their own separate requirements. The act is crucial to the nation's well-being in recognizing that the coast is where a battle for life on this continent will take place and in encouraging the citizenry to make this battle the nation's business. It is a battle that the nation, by itself and without excessive disruption, can win.

Once freed in its rivers, its great bays, its sand-swept shore, allowed to erode where it will, build where it will, to respond as it will to currents, to winds, to the sun, to moon-made gravity, the coast, deprived, stressed, over-burdened as it is, can begin to restore itself. To some extent it can resume its free-flowing character, its use of tidal energy to flush out estuaries, carrying marsh-manufactured food to inhabitants of the crucial ocean fraction offshore. It can reestablish the stunning diversity that provides habitats for the spectrum of coastal creatures now surviving, from migrating shorebirds to rock-based bivalves. It can once again build its beach-dune defenses against flood, hurricane, winter storm. It can again encourage flora and fauna that depend on the narrow strip where rivers deliver earth nutrients to mingle with those of the ocean, where fresh water meets salt, land meets sea.

And all this will provide the nation with a good supply of poison-free fish and shellfish, with sand beaches and water safe to swim in, with the matchless beauty of a working shore.

What is in question is an evolutionary leap such as no generation before us has attempted. It will require all the musculature of the human mind, flexed as it has been by the preparatory steps of the past few years, to strip off the entanglements of coastal use rooted in several centuries of settlement and development, and move to new man-coast relations in which we will acknowledge and concede the shore's requirements. We give up gambling on the outside limits of what the coast can take in order to discover what it can become with our meticulous stewardship.

Extend the linkage that inspires this great leap and it encounters the impact of other coasts *in extremis* beyond the horizon here and abroad, other people pouring their poisons into the ocean that impinges on all the coasts of the world, driving species to extinction and otherwise threatening the shore. The significant radical sacrifice of many Americans—giving up, forgoing, changing long-time expectations—to restore the coast to the safety of its optimum natural function and to keep it that way, could make us a force of greater importance than might have been foreseen, freeing all the improbable shore systems of this lone green and blue planet where the thin edge has, miraculously, come to pass.

A BIBLIOGRAPHICAL GUIDE

The following short guide to published information has been excerpted from the extensive source material consulted in the preparation of this book, which is to be found in the publications of certain public and private agencies, in newspapers, magazines for the general reader, in journals devoted to specific subject areas, and in a wide variety of books. The sources I have selected here are of particular interest and importance and can lead to a lifetime of coast-related reading.

U.S. government agencies

Army Corps of Engineers.
Council on Environmental Quality.
Department of Commerce, particularly its National Oceanic and Atmospheric Administration under which are significant offices—the Environmental Research Laboratories, Marine Eco-Systems Analysis Program,

National Marine Fisheries Service and the Office of Coastal Zone Management.

Department of Interior, particularly the Bureau of Outdoor Recreation, Fish and Wildlife Service, Geological Survey, National Park Service.

Environmental Protection Agency.

National Science Foundation.

Office of Technology Assessment.

Senate Committee on Interior and Insular Affairs.

Senate Committee on National Ocean Policy Study.

State and local government agencies

Each state can choose a "lead agency" to direct its coastal zone managment programs and these are listed in the CZM office in Washington. To discover other state agencies with coastal responsibility, and pertinent publications, query this "lead agency" in the state. Local publications come from a spectrum of official sources differing according to the size, sophistication and coastal involvement of town and county.

Private Agencies with newsletters and other useful publications

American Littoral Society.

The Conservation Foundation.

Environmental Defense Fund.

Ford Foundation.

Friends of the Earth.

Environmental Policy Center.

Massachusetts Institute of Technology's Center for Policy Alternatives.

Marine Technology Society.
National Audubon Society.
Natural Resources Defense Council.
The Nature Conservancy.
New England River Basins Commission.
Regional Plan Association of Greater New York.
The Research Institute of the Gulf of Maine.
The Sierra Club.
The Urban Land Institute.

Some often-used newspapers and magazines

Coastal Zone Managment
Coastal Zone Management Journal
Columbia Environmental Law Journal
Environmental Law
Maine Times
Marine Fish Management
National Parks and Conservation Magazine
North Carolina Law Review
Not Man Apart
New York Times
Oceans
Oceanus
Wall Street Journal
Washington Post

Selected Books

Ackerman, Bruce A. et al. *The Uncertain Search for Environmental Quality.* New York: The Free Press, 1974.
Albion, Robert G., William Baker and Benjamin Labaree. *New England and The Sea.* Middletown: Wesleyan University Press, 1972.

Audubon, Maria R. *Audubon and his Journals* (2 vols.). New York: Charles Scribner's Sons, 1897.

Baldwin, Pamela L. and Malcolm Baldwin. *Onshore Planning for Offshore Oil.* Washington, D.C.: The Conservation Foundation, 1975.

Baram, Michael S. *Environmental Law and the Siting of Facilities.* Cambridge, Mass.: Ballinger Publishing Company, 1976.

Bartram, John and William. (Helen Cruickshank, ed.) *John and William Bartram's America.* New York: The Devin-Adair Company, 1957.

Borghese, Elizabeth Mann. *The Drama of the Oceans.* New York: Harry N. Abrams, Inc., 1975.

Bradford, Peter Amory. *Fragile Structures: A Story of Oil Refineries, National Security and the Coast of Maine.* New York: Harper's Magazine Press, 1975.

Cain, A. J. *Animal Species and their Evolution.* London: Hutchison University Library, 1971 edition.

Calder, Nigel. *The Restless Earth.* New York: The Viking Press, 1972.

Carter, Governor Jimmy. *A Blueprint for Action: Goals for Georgia in the Seventies.* Atlanta: State of Georgia, n.d.

Clark, John. *Coastal Ecosystems: Ecological Considerations for Management of the Coastal Zone.* Washington: The Conservation Foundation, 1974

Comptroller General of the United States. *The Coastal Zone Management Program: An Uncertain Future.* Washington: U.S. General Accounting Office, December, 1976.

Cousteau, Jacques-Yves. *Life and Death in a Coral Sea.* Garden City: Doubleday and Co., 1971.

Culliney, John L. *The Forests of the Sea.* San Francisco: Sierra Club Books, 1976.

Dawkins, Richard. *The Selfish Gene.* New York and Oxford: Oxford University Press, 1976.

Dibner, Martin. *Seacoast Maine, People and Places.* Garden City: Doubleday and Co., 1973.

Ditton, Robert B., John L. Seymour and Gerald C. Swanson. *Coastal Resources Management.* Lexington, Massachusetts, Toronto: D. C. Health & Company, 1977.

Gibbons, Boyd. *Wye Island.* Baltimore and London: Johns Hopkins University Press, 1977.

Gorges Society. *Rosier's Relations of Waymouth's Voyage to the Coast of Maine, 1605.* Portland, Maine: The Gorges Society, 1887.

Grad, Frank P. *Treatise on Environmental Law* (revised). New York: Matthew Bender and Co., Inc., 1976.

Hanie, Robert (Kenneth Brower, ed.). *Guale, the Golden Coast of Georgia.* San Francisco, New York, London, Paris: Seabury Press, 1974.

Hay, John. *Spirit of Survival.* New York: E.P. Dutton and Co., Inc., 1974.

Ketchum, Bostwick H. (ed.). *The Water's Edge: Critical Problems of the Coastal Zone.* Cambridge, Mass. and London: The M.I.T. Press, 1972.

Liroff, Richard A. *A National Policy for the Environment— NEPA and its Aftermath.* Indiana University Press, 1976.

Lucas, Joseph and Pamela Critch. *Life in the Oceans.* New York: E.P. Dutton and Co. Inc., 1974.

Ma⋯ie, G. E. and Nettie MacGinitie. *Natural History ⋯ Anim⋯ls* (2nd Edition). New York: McGraw Hill, 1968.

Magnuson, Senator Warren G., (chairman). *The Oceans and National Economic Development.* Washington: U.S. Government Printing Office, 1973.

Matthiessen, Peter. *The Wind Birds*. New York: The Viking Press, 1973.

Milne, Louis J. and Margery Milne. *Ecology Out of Joint*. New York: Charles Scribner's Sons, 1977.

Mogulof, Melvin B. *Saving the Coast*. Lexington: D.C. Heath and Co., 1975.

Moorcroft, Colin. *Must the Seas Die?* Boston: Gambit, 1973.

Morison, Samuel Eliot. *The European Discovery of America* (2 vols.) New York: Oxford University Press, 1974.

Mostert, Noël. *Supership*. New York: Alfred A. Knopf, 1974.

Norman, J. R. and P. H. Greenwood. *A History of Fishes*. (3rd edition). New York: John Wiley and Sons, 1975.

Odum, Eugene P. *Fundamentals of Ecology* (3rd edition). Philadelphia, London, Toronto: W. B. Saunders Company, 1971.

Peterson, Roger Tory. *A Field Guide to the Birds*. Boston: Houghton Mifflin Company (1947 edition).

Sax, Joseph L. *Defending the Environment*. New York: Alfred A. Knopf, 1971.

Scott, Stanley. *Governing California's Coast*. Berkeley: University of California Institute of Governmental Studies, 1975.

Shepard, Francis P. *Geological Oceanography*. New York: Crane, Russak & Company, Inc., 1977.

Silverberg, Robert. *The World Within the Ocean Wave*. New York: Weybright and Talley, 1972.

Simon, Anne W. *No Island Is an Island*. New York: Doubleday and Co., 1973.

Simpson, George Gaylord. *The Meaning of Evolution*. New Haven: Yale University Press, 1950.

Stone, Christopher D. *Should Trees Have Standing?* Los Altos: William Kaufman, Inc., 1974.

Teal, John and Mildred Teal. *Life and Death of the Salt Marsh.* Boston, Toronto: Little Brown and Company, 1969.

Ursin, Michael J. *Life in and around the Salt Marshes.* New York: Thomas Y. Crowell Company, 1972.

Wayburn, Peggy. *Edge of Life.* San Francisco, New York: Sierra Club, 1972.

Weinberg, Steven. *The First Three Minutes.* New York: Basic Books, Inc., 1977.

Wenk, Edward Jr. *The Politics of the Ocean.* Seattle and London: University of Washington Press, 1972.

INDEX

New York *(cont'd)*
  Hudson River, 68, 95–96, 97, 98,
    136
  Jamaica Bay, 19, 140–42, 155
  Long Island, 15, 96, 98, 103–04,
    160
  National Park Service, property
    acquired by, 84, 87, 141–42;
    Gateway National Recreation
    Area, 89–90, 141–42
  offshore oil, 103–04
  Verrazzano's description of New
    York Harbor, 30
  water pollution: nuclear power
    plants, 98; PCB, dumping of,
    68, 95–96, 97, 136; sewage and
    waste, 9, 30, 38, 39, 89, 95,
    96–97, 98, 160
New York Times, 34, 111, 153
Nixon, Richard M., 81, 89–90, 109,
    110, 130
NOAA *see* National Oceanic and
    Atmospheric Administration
Noble Foundation, 82
Norman, J. R.: *A History of Fishes,*
    67
North Carolina: beach erosion, 15
  Cape Hatteras National Seashore
    Park, 84
  coastal birds, 65
  wetlands, loss of, 51–52
Northern Gulf oil spill, 102
Norway: offshore oil, 105–06, 112
NPS *see* National Park Service
nuclear power plants *see* power
    plants/nuclear power plants

Ocean City (Md.), 86
OCS *see* Outer Continental Shelf
Odum, Eugene, 35
  Fundamentals of Ecology, 21
  marsh grass and marshes, study
    of, 46, 49–50, 78
Office of Technology Assessment,
    99
Offshore Power Systems, 93
Ogburn, Charlton: *The Winter Beach,*
    16
Oglethorpe, Gen. James, 73
Ogunquit (Me.), 127

oil: American Petroleum Institute,
    101, 135
  Alaska, 107, 153
  coastal areas affected by, 99–112
  Geological Survey, 103–04, 106,
    111
  Great Britain, 101, 149, 155
  Japan, 111
  leakage and spillage (and effects
    of), 6, 11, 27, 37, 38–39, 60,
    68, 78, 100–06 *passim,* 154
  legislation, 108–09, 128, 161;
    Coastal Zone Management,
    131, 135–36, 137
  Norway, 105–06, 112
  offshore exploration and drilling,
    6, 10, 11, 26, 91, 92, 93,
    99–100, 103–12, 121, 135, 137,
    149, 151, 153–54, 155, 161
  refineries, 6, 53, 55, 56, 93, 111,
    122, 161
  tankers, 34, 161; ports, 6, 27, 34,
    55, 122, 128; *see also* leakage
    and spillage *above*
  U.S. reserves and use, 106–07
Oregon: Coastal Zone
    Management, 133; Coos Bay
    estuarine sanctuary, 131
Ossabaw Island (Ga.), 74, 80–81,
    82, 83, 148, 160; *see also*
    Georgia, barrier islands
otters, 5, 43
Outer Continental Shelf (OCS): oil
    exploration, 6, 92, 99–100,
    103, 104, 106–11 *passim,* 121,
    137, 151, 153–54, 161; *see also*
    Baltimore Canyon; Georges
    Bank
oysters, 4, 13, 42, 54, 55, 58, 59
  Delaware Bay, 53–54, 55
  depletion and decline, 53, 55, 61,
    79
  in estuarine system, 44, 46, 47,
    53, 54, 55
  oil spill, effect of, 102
  water pollution, effect of, 54–55,
    98

Palm Coast (Fla.), 81–82, 84
Parsons, George, 74, 82